The Unintelligent Designer

Refuting the Intelligent Design Hoax

Rosa Rubicondior

The Unintelligent Designer

Cover photograph:
  Emerald cockroach wasp (*Ampulex compressa*).
  Internet source.

18 Sep 2018 – Minor revisions and corrections.

14 Jun 2021 – Minor revisions and corrections.

Third Party Copyright.

ISBN-13: 978-1723144219
ISBN-10: 1723144215

"If a theory claims to be able to explain some phenomenon, but does not generate even an attempt at an explanation, then it should be banished. Despite comparing sequences and mathematical modelling, molecular evolution has never addressed the question of how complex structures came to be."

Michael J Behe, *Darwin's Black Box*, p.185 (1996)

[Mr Eric Rothschild (for the plaintiffs) interrogating Michael J Behe (witness for the defence)]

ER. Now you have never argued for intelligent design in a peer reviewed scientific journal, correct?

MJB. No, I argued for it in my book.

ER. Not in a peer reviewed scientific journal?

MJB. That's correct.

ER And, in fact, **there are no peer reviewed articles by anyone advocating for intelligent design supported by pertinent experiments or calculations which provide detailed rigorous accounts of how intelligent design of any biological system occurred**, is that correct?

MJB. **That is correct, yes.**

Tammy Kitzmiller, et al. v. Dover Area School District, et al. 2005
United States District Court for the Middle District of Pennsylvania
Day 12 (October 19), AM Session, Part 1

The Unintelligent Designer

Dedication

I dedicate this book to the biologists and palaeontologists who work through the scientific method, to discover the truth and who, in doing so, quite incidentally and without effort or intent, continue to falsify creationism and its recent, pseudo–scientific, lab–coated version, Intelligent Design.

It is dedicated also to the army of informed individuals who effectively combat the well–financed propaganda efforts of those politically–motivated creationists and their willing dupes in the social media and who do so much to inform the casual observer about the truth of science.

Probably due to the efforts of these individuals and, it has to be said, the ineptitude and intellectual dishonesty of the Creation Industry and its dupes, the USA recently saw a fall in support for creationism to below 40% for the first time since polling on this question began in the 1981. The fall was only to 38%. Although this is still an appalling level of scientific ignorance it is none–the–less welcome especially since the largest fall was seen in the younger adults.

This book is dedicated especially to the people responsible for this welcome success.

The Unintelligent Designer

Contents

The Unintelligent Designer

Introduction

'Intelligent Design' (ID) is a recent invention of the American evangelical conservative Christian right, invented with the intention of presenting creationism as real science, at least in the minds of people who don't understand science. The purpose is to persuade legislators that, as a science, not a religion, it should not be barred by the First Amendment, specifically the Establishment Clause, of the US Constitution from being taught at public expense in public (i.e., publicly funded) schools.

The First Amendment states:

> Congress shall make no law respecting an establishment of religion, or prohibiting the free exercise thereof; or abridging the freedom of speech, or of the press; or the right of the people peaceably to assemble, and to petition the Government for a redress of grievances.

The first clause of this is commonly referred to as the Establishment Clause. Several court cases [1] have established that creationism is a religious opinion, not a scientific fact, and therefore comes within the scope of the First Amendment. Obviously, this would not apply if legislators could be persuaded that 'Intelligent Design' is simply an alternative scientific opinion on a par with differences of opinion about why dinosaurs became extinct or what exactly happened in the Big Bang, derived from an examination of the same set of data as the view that life evolves on Earth.

The ID lobby seeks to appeal to those with little or no scientific knowledge or understanding of science and the scientific method, and is thus able to ignore major deficits in the notion and simply appeal to ignorant incredulity and deep–seated cultural parochial arrogance and chauvinism. They operate in an anti–intellectual culture which is highly suspicious of science and 'elitist' scientists, viewing learning with suspicion and academic achievement as getting above one's station in life.

Ironically, the ID lobby benefits too from a culture which views government itself, although democratically elected, with suspicion and the assumption that the government is the enemy of the people. Government is believed to have a hidden agenda which includes imposing some sort of 'Liberal', 'Socialist', even 'Communist' agenda on a conservative people, waging a war on Christianity, working for big business and the agrichemical and pharmaceutical industries, and of course, in some extreme forms, being part of an international Jewish conspiracy run be a secret committee based somewhere in the little–understood and vaguely sinister place called 'Abroad'.

In this culture, the US Constitution itself is seen as anti-Christian, or at least wrongly interpreted. The 'Founding Fathers', despite all the evidence to the contrary and despite Thomas Jefferson's insistence on a 'wall of separation between church and state', are believed to have intended to found the nation on solidly fundamentalist Christian principles with a Declaration of Independence probably written by God himself and a Constitution based on the Bible, continuing the supposed aim of the early 'Pilgrim Fathers' to build a New Jerusalem. According to this contra–historical view, God himself created America for his chosen people – American fundamentalist Christians – and every day God blesses America.

Overthrowing the First Amendment is thus an objective in itself of many conservative Christian organisations and of the Discovery Institute, which is largely responsible for the ID movement. The aim being to create the America that God intended – a fundamentalist, Protestant Christian theocracy, with laws based on the Ten Commandments and Leviticus and the Bible taught as science and history. To take America (and then the rest of the world) back to some assumed golden age in the Late Bronze/Early Iron Age.

Introduction

Although many Christians fully accept evolution as scientifically established, with Genesis as some sort of allegory or metaphor, and so see no conflict between their faith and science, Bible literalist fundamentalists see science as a threat, refuting as it does almost every line in Genesis as literally interpreted. The ID movement more than most Christians, realises that if the science is correct, Genesis is not history, and if Genesis is fiction, the entire premise of the Bible as the inspired word of an omniscience, honest creator god is unsustainable, as is the notion of a first couple, original sin and the need for Jesus as a saviour. And if that goes, the entire Christian religion, together with the power and control it gives to the priesthood goes with it.

As I will show, ID is unsustainable as a scientific notion (I won't grace it with the term 'hypothesis' or 'theory' – terms which have specific meanings in science, none of which can be applied to ID).

When examined in anything more than a superficial level, living things show no evidence of design, other than that of design by a natural, mindless, amoral and undirected process which can be neither described as intelligent nor design, and certainly not intelligent design by anything remotely resembling the god of the Bible and Qur'an. Let us not forget that creationism is not confined to fundamentalist Christianity but is also preached by mainstream Islamic imams, Mormons, Sikhs and some Hindu gurus with each trying to force–fit observable reality into their preferred origin myths most of which go back to what Christopher Hitchens called, the fearful infancy of our species.

In the following chapters I will give examples, backed up by recent research, that shows how living systems, including ecosystems and interdependent and parasite–host systems cannot possibly be described as designed, nor any designer of such a system described as intelligent and most certainly not loving and compassionate.

I will also look at the design process itself. As we will see, there is nothing in living systems that resemble what the outcome from a good, intelligent design process would produce, especially if the designer was omnipotent and so capable of ultimate perfection of design.

Some of the examples I have used in the book are so highly unlikely that I've included a bibliography so you can read about them for yourself just in case you are tempted to think I must have made them up to discredit this supposedly intelligent designer.

All of the examples are examples of multiple flaws in the ID argument so it is to an extent arbitrary whether to use them as examples of needless complexity, prolific waste, arms races, etc. For example malaria is an example of needless complexity if the purpose is just to make more malaria parasites and of prolific waste in the astronomical numbers of parasites produced just to ensure a few make it to the next generation. It is also an example of an arms race as the parasite becomes resistant to drugs used to treat it and the sheer malevolence of any designer who would design such a thing and go to such lengths to make people sick and kill so many.

Some of the material is based on articles previously published in my Rosa Rubicondior blog, http://rosarubicondior.blogspot.co.uk.

As with all my books and blog articles, never take what I have said as definitive truth without checking yourself. I believe all the facts to be true but I don't want you to believe me; I want you to believe the evidence.

1. What Is Design?

Any argument about evidence for design in living things must first agree on a definition of design against which to compare the systems under consideration. Having been involved in debates (I use the word lightly) in which creationists have tried to define Intelligent Design as anything designed by God and therefore encompassing the entirety of 'creation', I want to avoid the obvious and intellectually dishonest circularity of such arguments. I will thus use the definition used by people and organisations who know about the principles of design but who have no religious or political agenda.

The former Chairman of the UK Design Council, Sir George Cox, said:

> Design is what links creativity and innovation. It shapes ideas to become practical and attractive propositions for users or customers. Design may be described as creativity deployed to a specific end [2].

The operative words here being 'practical' and 'specific end'. Design is practical and has to suit the specific needs of the user or client. In other words, the thing being designed must have a clear purpose.

As an information systems manager one of my responsibilities was to produce software solutions to meet the information needs of managers and directors of a UK NHS Trust. Whatever the finished solution was, the essential first step was always to get a very clear understanding of the end users' needs. In other words, to get a clear and agreed objective; what was the solution expected to do, exactly. What was its specific purpose? Only then could we begin to design solutions to meet those exact requirements.

The Unintelligent Designer

In an article on good design principles for information technology, addressing the complexity vs simplicity question, design expert John Spacey said:

> Complexity versus simplicity is a common design tradeoff. Complexity always has a cost. As such, complexity is ideally minimized for equivalent functionality and quality... Adding complexity without adding functionality or quality is known as needless complexity. Complexity can be exciting and it is possible to get involved in making technologies, communications or ideas complex for the sake of complexity. Generally, this is a mistake as complexity costs more to develop, support and use [3].

It is really common sense that a hallmark of good design is minimal complexity and maximal simplicity. If the same objective can be achieved with a simple design as with a complex design, then what is the purpose of complexity? Complexity not only carries a production cost but also an operating cost in that there is more to go wrong. Complex solutions are usually more difficult to use than simple ones.

Needless complexity is poor design, not good design. It is not evidence of perfect design but of imperfect design; not of intelligent design but of stupid design.

Take for example, the gardeners' tool, the simple dibber or dibble, used for making holes in soil for seed, bulbs and plants. How can a simple, tapered piece of wood, maybe with a scale engraved on it to measure the depth of the hole for different seeds or bulbs, be improved on? How could it be made more efficient to use and maintain, and cheaper to manufacture by adding complexity?

It can't be, of course. Any added complexity would be unnecessary and detrimental; a reduction in 'design', not an improvement. Why would any intelligent designer design a garden tool that any intelligent gardener would reject as too expensive, too complicated to use and too difficult to maintain and clean, in favour of a cheaper, more useful, simpler and easier to maintain and clean tool for doing the same thing?

In summary then, we can say the hallmarks of good, intelligent design are:

- Clear functionality, so that the purpose of the design is clear to see and meets the requirements.
- Maximal simplicity and minimal complexity.

The latter requirement especially, is particularly relevant when we come to look at biological systems, especially since ID advocates repeatedly cite complexity as evidence of intelligent design.

This appeal to the 'obviousness' that complex design requires intelligence is nothing more than an appeal to ignorance or lack of imagination. It is an appeal to intellectual laziness and the arrogance of ignorant incredulity.

Take, for example, graded pebbles on a beach or the little ripples which form in sand in response to wind on sand dunes or flowing water on a beach, or the myriads of different snowflakes, and compare them as examples of design to the humble garden dibber I described earlier. Which of those is more complex? Which of those were 'designed' by the simple operation of unintelligent natural forces – of chemistry and physics – and which was designed by an intelligent designer?

That question is rhetorical, of course. In fact, the 'obvious' answer is that the simple design is intelligently designed while the complex ones were not. The other, slightly more subtle consideration is that the examples of natural design have no obvious purpose! There is no obvious purpose in graded pebbles or ripples in sand and no obvious purpose in the 'design' of a snowflake. Indeed, if the design of a snowflake had any purpose, why are there so many different ones?

The Unintelligent Designer

As another example of unintelligent design, I will quote from an article I was inspired to write a few years ago following a visit to beach in Tunisia [4]. It is also an example of how order can emerge from chaos under nothing more than the operation of natural forces.

> Some years ago, on our first visit to the beach in Port El Kantaoui, Tunisia, we were astonished to find it littered with suspicious-looking, slightly flattened fibrous balls, especially along the tide line. Were these washed-up camel turds, maybe? Should we complain to the hotel on whose private beach all this stuff was scattered? Was this beach fit to walk barefoot on?

> Eventually, curiosity got the better of us and I gingerly picked one up. It was dry and when I sniffed it (one has to be prepared to make sacrifices for science) it was odourless save for a slightly salty sea smell. It was clearly made of tightly matted plant fibres which could be pulled apart. Maybe the balls were seed cases of some sort but we could find no seeds in them and what sort of seed case would be made of randomly arranged fibres? You would expect there to be some sort of structure within the ball itself, but there was none.

> Being an inveterate collector of natural curiosities, I simply had to take some home. I still have one in my little museum alongside fossil trilobites, a plesiosaur vertebra, pieces of fossilised wood, fossilised coral, sea urchins and ammonites - the latter three picked up from a single field in Buckinghamshire, England.

> But still we had no idea what these balls were or if they really had passed through the digestive system of a camel, or a fish of some sort. And why did they come in assorted sizes from one or two inches to some six inches or more. Did they all start off the same size and get worn down by wave action? So many questions; so few answers. The only thing for sure was that

they had come from the sea and had been washed up and left stranded along the tide line.

The basic problem was that they were clearly made of plant matter and had shape and form, but they just didn't look like they had grown and didn't seem to have any purpose.

The problem was that we were looking for evidence of natural design with a function, because that is what we normally see in natural things. Evolution creates the appearance of design and designed things have a function. A seed case is attached to the plant by a stalk (there was no sign of a stalk) and has seeds in it (there were no seeds). But these balls of randomly arranged plant fibres had form but no obvious purpose since they were clearly not seed-cases and were not vegetative dispersal phases like tumbleweed. Clearly they had not been designed, yet they were not mere random assemblages, although the individual fibres were randomly arranged.

What I was looking at was structure which had emerged from chaos.

I now know that these fibre balls, called Poseidon balls, are common on Mediterranean beaches. If you've been to the North African coast, or the beaches of Greek Islands, you will very probably have seen them. They are produced by wave action on the dead fronds of the seagrass *Posidonia oceanica* or rather the tough 'veins' which are left behind when the softer parts decay. *P. oceanica* is still relatively common in the Mediterranean where it is an important component of the ecosystem, but it is under threat from pollution. Curiously it grows nowhere else apart from the sea around Australia.

This seagrass grows offshore, just beyond the breakers and the fibres clump together under the action of waves and tidal currents, sometimes into long sausage-shaped rolls which then break into sections. The clumps are picked up by waves and, like pebbles, are rolled around and worn into the shapes we see deposited along the tideline.

There is no direction to this process save that provided by the slope of the sea-bed. At all stages everything is random, from the movement of water molecules to the alignment and distribution of the fibres, yet the shape of the seabed, which causes the random waves to break in a definite zone, then the random movement of the forming balls, produces structure and form and quite definitely a non-random structure at the level we see laying on the beach. If William Paley had found one, how would he have fitted this into his watchmaker model of a designed universe?

Despite what creationists tell you, order can come from chaos under nothing more than a directional force (in this case the slope of the sea bed). In the case of the formation of galaxies and stars the direction is provided by gravity; in the case of living organisms the direction is cause by non-random natural selection which pushes randomness towards 'design' for survival and the function of reproduction.

No intelligence, no plan, no magic and no magic creators required.

So, we can see then that purpose (as distinct from function) and simplicity are what we would expect to see in something that has been designed, and especially something designed by an intelligent, perfect entity.

We will see that nothing in nature even approaches that.

It should be clear then that the superficial appearance of design is misleading. When compared to something that **has** been designed, living systems and ecosystems have none of the characteristics of good design.

When examined in detail we find many examples of poor design, one of which is an astonishing level of needless complexity. Another of which is evidence of competition between species with one species' solution being another species' problem to be solved. In other words, we see evidence of stupidly wasteful competition and arms races, none of which can possibly be described as intelligent design. Chapter 3 examines these arms races; first I will deal with the argument from design, or the Teleological Argument.

2. The Argument from Design

The entire intelligent design argument is essentially a rehash of the old argument from design or Teleological Argument, popularised by Reverend William Paley in the early 19th Century, so it seems appropriate to get this generalised argument out of the way first of all. It was devised at a time when the minutiae of biological detail were unknown because we lacked the tools and the science to examine living things in that detail.

The modern ID movement has simply taken Rev. Paley's argument and applied it to the detail, but the same refutation applies equally well at that level as it did at the level at which Rev. Paley employed it.

The following is based on an article I wrote in 2012 [5] and which I first published in book form in *The Light of Reason Vol II: Atheism, Science and Evolution* [6].

The argument from design or Teleological Argument has a long history, but perhaps its most famous exponent was William Paley, the early 19th century English theologian and philosopher. Briefly, his argument, which pre-dated Darwin's Origin of Species by 57 years, was that, if you were walking along a heathland path and found a watch, the most logical conclusion would be that someone had dropped it there and that it had been designed and created by one or more watchmakers and not by natural forces. It would be irrational to conclude that it had spontaneously self–assembled.

Of course, this is unarguable for a watch, for the simple reason that there is no other mechanism which could explain the watch's production, nor how it came to be where it was found. That explanation requires no mystery; there is nothing required which can't be readily understood and certainly there is no need to include an unproven supernatural hypothesis in the explanation. The explanation that a watch was designed and made by a watchmaker is complete and the most parsimonious answer available.

With the state of our knowledge of biology and biological systems in 1802, there seemed to be no reason why this analogy did not apply to living animals as well. Living animals appear to be designed in that they have component parts which need to be arranged in the right way; if any of the major components are missing the whole will not work.

But, what purpose does a living animal have which is in any way comparable or analogous to the utility value of a watch? Living things appear to exist only to produce other living things but this is by no means true for watches. Watches have a very specific purpose; to keep an accurate record of the passage of time for its owner.

Watches are unlike living things in an even more fundamental way; watches do not need elaborate mechanisms for finding their own energy source and to avoid becoming some other timepiece's energy source; Because they do not eat and drink they do not need excretory and circulatory systems to supply energy to its component parts and to carry away the waste, and, most significantly, watches do not need mechanism for finding mates and for producing and caring for offspring.

Because they do not need any of these things they do not need sensory, reproductive and locomotory systems and because they are not self-replicating, they need no mechanism for replicating information and passing it on to the next generation.

They don't need any of these things because they are designed and made by humans, for humans and humans provide their energy to them by winding them up (or in modern times, putting a new battery in them). Without humans, watches have no purpose, no function, and no existence. Watches are merely human artefacts. Living creatures existed before humans and would undoubtedly exist without us. For the most part, living creatures are self-reliant and self-replicating because they have no designers and have no purpose other than existing for their own sake.

Look inside a watch you will not find any redundancy in the design. What you'll find is economy and minimal complexity. There will be no cogs spinning purposelessly away, no springs holding back levers for no reason at all, no overly elaborate mechanisms using several cogs and levers where one or two would do, no mistakes having to be compensated for by hugely inefficient workarounds and no evidence of earlier designs still included but having no current function at all.

The watch would be efficiently and accurately designed with obvious intelligence by someone who had a complete over-view of the purpose of his design and who knew how to make it as simply and therefore as efficiently and accurately as possible, adding only such decoration as pleases the owner and makes a sale more likely.

Additionally, if you were to look in different models of watch made by this watchmaker you would certainly see the same solutions used to overcome the same engineering problems; you would see the same springs, cogs, levers and bearings being used in the same way. You might even see exactly the same mechanism, just in a different case.

Unlike watches, living things have masses of inbuilt redundancy and needless complexity. The DNA of most living things is vastly more than is needed. There is DNA which does nothing other than produce copies of itself. There is DNA which is added to the ends of chromosomes for no good reason because of a flaw in the copying mechanism and which just keeps being added to. There are the remnants of ancient retroviruses, so going back to our fish ancestry or earlier.

There are vestigial organs to be found in most species, like evidence of legs in whales and the human appendix. As we will see in Chapter 5 with the example of *Trichomonas vaginalis*, with 27 of the 29 genes needed for meiosis and sexual reproduction even though it never reproduces sexually and never undergoes meiosis; humans and other primates have the first three enzymes for manufacturing vitamin C all fully functional yet the fourth enzyme is broken by a simple mutation, so we never manufacture our own vitamin C like many other creatures do [7; 8]. There is massive waste and atavism. There is evidence of workarounds for earlier mistakes such as a complicated neural function to compensate for the blind spot in the mammalian eye because the wiring of the retina is backwards. And of course there is the ludicrous path taken by the recurrent laryngeal nerve, especially in the giraffe [9].

There is evidence of repeated new 'designs' of structures like wings and eyes and not the re-use of earlier solutions, such as a watchmaker would use. No intelligent watchmaker would think to re-design springs and cogs each time he decided to make a new watch.

In short, living things show evidence of design, but not of intelligent design.

So, does that apparent design point to a god, but just not a very intelligent one perhaps, or one with a fixation with beetles, of which there are some 500,000 different 'designs' alone?

What Paley, and those who were convinced by his argument, which incidentally included a young Charles Darwin, did not appreciate, in addition to all the redundancy, and in addition to failing to appreciate that watches have an obvious purpose which is not paralleled by living things, was that design does not necessarily indicate a designer, nor intelligence. This was never more than an argument from personal incredulity - I can't understand it therefore it must have been a god (and of course the locally popular god is the only one in contention; the god that Rev. Paley was actually proselytizing for). They failed to appreciate this not because they were stupid or dishonest; they could only work with the state of knowledge of the times. They failed to appreciate it because they lacked one essential piece of knowledge, because science had not discovered it then.

The Argument from Design

What they failed to appreciate was that there is a natural process which can explain **all** these things, and which does not include an unexplained mystery for which no hypothesis can account, nor does it require magic. All the components of this system can be seen and understood, just like all of the components of the system for making watches can be seen and understood. No mystery, no magic and no supernatural component need be included in the explanation. Given the system in which natural selection operates, the result is inevitable.

Natural selection is the most parsimonious explanation both for the appearance of design and for the appearance of a stupid designer. Living things look exactly as you would expect them to look if designed by a utilitarian, mindless, purposeless design process given direction only by the environment in which it operates.

Now that we can stand on the shoulders of giants like Darwin and Wallace, we can see further than other men. We can now see further than the Bronze Age goat-herder who thought up the creation myth and who couldn't even see over the horizon – and who thought the earth was flat.

We can see now that there is nothing supernatural required and nothing supernatural involved, and we need pay no heed to the ignorant gibberings of superstitious simpletons who insist it was all the work of their own small gods, and the clamour of the parasitic charlatans who feed off their ignorance.

The Unintelligent Designer

3. Arms Races

Biological arms races are the underlying forces that shape many of the relationships between species in an ecosystem, especially in a predator–prey relationship. It can take many forms and includes parasite–host relationships which will be the subject of the next chapter. This chapter concerns other arms races.

It is a matter of semantics whether a predator–prey relationship is a parasite–host relationship so I have defined a parasite as something which lives physically on or in its host and conversely, a host as something on or in which a parasite lives. To nature, of course, a host is simply a resource in the parasite's environment; a niche to be occupied, just as much as a herd of wildebeests is a host on which a pride of parasitic lions lives.

Remember, in the ID model, especially those advanced by Christian and Islamic fundamentalists, there is only one designer at work. American ID advocates have to be careful not to mention the Bible or the god described in it because of the fiction that ID is science not religion. Nevertheless, you will never see a Christian or Islamic ID advocate ever advance the notion that there could be millions, even billions of intelligent designers at work, all competing with one another and never sharing their solutions to design problems.

And yet, in biological arms races what we see is evidence that whatever is causing the many arms races, it looks like 'designers' working in complete isolation from each other and working assiduously to overcome the problems the other designers have designing as solutions to competitor's designs. Why would any intelligent designer design problems for itself in the form of solutions to the problem it designed as a solution to something else's problem – that it designed earlier?

The Berkeley University website, Understanding Evolution, gives this general description of an arms race between plants and herbivores. Plants have an interest in not being eaten while herbivores have an interest in eating plants:

> Consider a system of plant-eating insects. Any plant that happens to evolve a chemical that is repellent or harmful to insects will be favored. But the spread of this gene will put pressure on the insect population — and any insect that happens to have the ability to overcome this defense will be favored. This, in turn, puts pressure on the plant population, and any plant that evolves a stronger chemical defense will be favored. This, in turn, puts more pressure on the insect population...and so on. The levels of defense and counter-defense will continue to escalate, without either side "winning." Hence, it is called an arms race. This sort of evolutionary arms race is probably relatively common for many plant/herbivore systems [10].

Caterpillars and Brassicas.

A very nice example of this in practice was discovered in 2015 and published in the Proceedings of the National Academy of Sciences (PNAS) [11]. It made the popular press because it showed how evolution had given us mustard, as well as different varieties of the brassica family, due to an arms race between them and butterflies. An international research team led by University of Missouri Bond Life Sciences Center researchers, mapped the genes of the cabbage family onto those of butterflies and found that every change in the genes which produce a family of chemicals known as glucosinolates was followed soon after by changes in the butterflies [12].

Glucosinolates are responsible for the tangy flavours of mustards but they are distasteful and toxic to caterpillars – unless the caterpillars can detoxify them. What this research showed was that, starting near the end of the Cretaceous, the so–called K–T Boundary, when a meteor strike caused a catastrophic climate change that caused a mass–extinction that exterminated many of the dinosaurs and other species, the ancestor of the brassicas began producing these toxic compounds.

The team identified a series of three waves of change across fourteen species of brassicales with similar waves of change across nine species of related butterflies, the Pieridae butterflies which includes the Cabbage white butterflies.

One of the problems with these arms races where there are several predators is that the prey species needs to retain a defence against the other species while responding to one that has evolved a way round its defences. This is accomplished by gene duplication, so the original is retained but the duplicated gene is free to mutate and is available for natural selection – in this case the newly–evolved predator now finding the new compound toxic or distasteful. Very often, only a small change in the compound may be necessary to overcome the caterpillar's defences. In this lovely little example of co–evolution, the evolutionary pressure now switches to the caterpillar which, as the new gene spreads through the population of brassicas, faces a reduced food resource.

Incidentally, creationists will tell you that no new information can arise by mutation, and that any mutation must be detrimental, conveniently 'forgetting' gene duplication as a source of new genes while retaining the old ones.

To go back to my original point; how can this system be described as the work of a designer, let alone an intelligent one? Repeatedly having to find work–arounds for problems created as solutions to other problems it created earlier is the antithesis of intelligent action. It looks exactly like two species adapting to environmental change by genetic evolution without any planning or ultimate objective and simply building layers of increasing complexity in order to stand still, to the ultimate benefit of neither and with increasing costs to both.

Snake Venom.

Another example of an evolutionary arms race involving toxins is that of venomous snakes. This time however, the toxin is not produced as a defence against a predator but as a means of attacking a prey species. Once again, gene duplication is heavily involved and some proteins originally evolved for unrelated purposes have been adapted and utilised for new functions.

In 2013, an international team led by Freek J. Vonk of the University of Leiden, Netherlands published the results of a comparison of the genomes of the venomous Indonesian King cobra (*Ophiophagus hannah*) with that of the non–venomous constrictor, the Burmese Python (*Python molurus bivittatus*) and an anole lizard [13].

What they found was that the King cobra had assembled a large array of venoms, most of which were modified versions of proteins found in most cells and which perform basic metabolic processes there. The genes responsible for these fundamental metabolic proteins were common to all three reptiles but in the Cobra, repeated gene duplication had made them available for mutation and natural selection to enable the Cobra to overcome any resistance to its venom that had evolved in its prey whilst still being able to kill other, non–resistant prey species. The result of this evolutionary arms race is a cocktail of venoms, some of which are produce by the main venom glands and some by the accessory glands.

As an incidental aside, but pertinent to the subject of this book, was the finding that the Cobra has all the hox genes needed to build limbs except one – *Hoxd12*. It seems that the ancestor of the legless snakes lost the ability to grow legs by breaking or losing just one of the hox genes. Not only that, but the 'grow a leg' hox genes are still active in the snake embryo; they are just ignored by the cells. How can this be the result of intelligent design? What intelligent designer would provide almost all the equipment necessary to grow legs in a legless reptile and even make part of it work, all to no avail?

For more on this subject, see the *New Scientist* article by Bob Holmes of 4 June 2014 [14] and my blog post on the subject [15].

Tiger Snakes.

Remaining on the subject of venomous snakes and arms races for a moment; how evolutionary arms races can sometimes be won was revealed in a fascinating 2017 study of how the Australian Tiger snake venom works [16]. The study also answered the question why the same antivenom is effective against the venom of several Australian snakes. This appears to run counter to the idea that snakes speciate by evolving new venoms effective against new prey species.

The scientists found that Tiger snake venom blocks the blood–clotting mechanism at a level which is common to just about all animals. This mechanism is highly conserved in animals because any change would break the blood–clotting mechanism. This shows how an arms race was effectively ended because the predator hit its prey's Achilles heel and the prey species have never found a way round it.

Reindeer and Red Fescue Grass.

Maybe that is enough about snakes, at least for some readers. The next is another example of plants trying to defend themselves from herbivores and herbivores trying to overcome their defences, only this time it is slightly more complex than the one we met earlier with cabbages and caterpillars. It involves a third species; a fungus which is in a symbiotic alliance with a grass. Symbiosis is the process where two or more different organisms live in close association, each getting benefit from the association.

The widely–distributed grass, Red fescue, has formed a mutually beneficial alliance with the fungus *Epichloë festucae* which produces a highly toxic alkaloid, ergovaline. This can cause, amongst other things, Reindeers' hooved to fall off. Additionally, grazing the grass causes the grass to stimulate the fungus to grow, so producing more toxins.

Clearly, it would benefit herbivores in some way if they could inhibit this fungus and inhibit the signalling between the grass and the fungus which stimulates the fungus to grow. This is exactly what three scientists found Moose and Reindeer saliva does [17]. They showed that Moose saliva reduces the amount of toxin by between 41 percent and 70 percent but it also inhibits that growth–stimulating signalling.

If the ID lobby are to be believed, we have to accept that an intelligent designer would design grass, then design a herbivore to eat the grass, then set up a complex relationship between the grass and a fungus to help **stop** the herbivore from eating the grass, but one which depends partly on the grass being eaten to stimulate the fungus, and then would redesign the herbivore's saliva so it prevented that solution from working.

It would be a bizarre definition of 'intelligent' to describe such as designer as intelligent. It would be a bizarre definition of 'design' to describe such a complex and ultimately futile system as designed.

Another study [18] showed that Sheep saliva can actually stimulate a grass, *Leymus chinensis* to grow. It does this by stimulating the mobilisation of carbohydrates within the grass. Perhaps this is not an arms race but it is an interesting example of mutualism evolving. Could an intelligent designer, especially an omnipotent one, not come up with a better, simpler way to stimulate grass to grow so there was enough for Sheep to feed on?

Bats and Moths.

The following is an example of an arms race between predator and prey species; in this case between moth–eating bats and moths. It is based in large part on a very readable account by David Jacobs, Professor of Animal Evolution & Systematics, University of Cape Town writing in online magazine, *The Conversation UK* [19].

These competitive relationships are easy to predict from the basic principles of evolution because very obviously the animal that catches most food is more likely to leave descendants with their inherited abilities than one which doesn't have them or in which they are less efficient. Conversely, the individuals of the prey species which avoid getting eaten are also more likely to leave more descendants which will inherit their avoidance abilities than ones less good at avoidance.

It sometime seems so obvious that I hesitate to spell it out, except that there are creationists who profess to be baffled by how natural selection works to increase advantageous genes in the species genepool at the expense of the less advantageous alleles with which they are, in Richard Darwin's words, in deadly rivalry.

The arms races between bats and moths probably started when bats evolved echolocation as a way to 'see' in the dark. This would have made night–flying moths especially vulnerable. In fact, moths may have evolved a nocturnal habit to avoid being eaten by diurnal predators in the first place.

This then put evolutionary pressure on moths to be able to detect the sonar signals of bats, especially when they 'locked on', and take avoiding action of some sort. Bat echolocation call frequencies (12– 210 kHz) are normally above the hearing range of most humans who have an upper limit of about 20 kHz (probably fortunately in some areas where the night would be a very noisy place if we could hear the bats). A frequency of 1 kiloHertz (kHz) is a thousand per second; in the case of bat echolocation clicks, a thousand clicks per second.

Significantly, however, those moths which can detect bats have a hearing range at the same frequency as the bats that prey on them. Significantly too, moths don't appear to use this ability to hear for any other purpose. They don't signal to one another with sound, so its sole purpose as a sense seems to be to detect bats.

One avoiding action seems to be to emit clicks which deter the bats in some way. This can be observed when bats break off an attack in response to these clicks although it is not yet clear why – it maybe that they are signalling an unpalatable taste, which is normally signalled by coloration; a method not available in the dark. Are these clicks emitted by some moths in response to an impending bat attack the sonar equivalent of warning colours? That remains to be proven, but they can be observed to work. Another possibility is that the moth's clicks block the bat's echolocation or confuse the bat in some way.

The ability of moths to detect bats and take evasive action then created a problem and evolutionary pressure for bats to hide or disguise their presence. One way of doing this is to change the frequencies of their sonar so they are above or below the hearing range of their prey moths.

Evidence of this adjustment of both predator and prey hearing and echolocation frequencies is provided by regional variation. In North America, most insect eating bats have an echolocation frequency of 20–50 kHz which is about the same frequency at which moths hear well. In Africa and Australia, bats use frequencies above 50 kHz and moths can hear at frequencies up to 100 kHz.

Two bats have adopted opposite strategies; one, the North American Spotted bat, uses very low frequency calls (12 kHz) while the African Short–eared trident bat uses very high frequencies (208 kHz). Both live exclusively on moths. Both the Horseshoe bats and the Old World Leaf–nosed bats also use very high frequencies and feed predominantly on moths

But there are signs that in some places moths are ahead in the arms race whilst in others the bats currently have the upper hand. In the south coast area of South Africa for example, the Cape Horseshoe bat's frequency of 80–86 kHz are audible to most common moths in the area, but the Dusky Leaf–nosed bat's calls are inaudible to most moths.

Another way in which bats can make themselves undetectable to moths, at least until it is too late for the moth to take evasive action, is to change not the frequency of their calls but the loudness of them. A moth's ability to detect a call is a function not just of the frequency but of its amplitude or loudness. One bat, the European Barbastelle bat uses a call intensity of 10 to 100 times lower than other bats in the same area and targeting the same species. The success of this strategy can be seen from the fact that the Barbastelle feeds almost exclusively on eared moths.

Here again we have an arms race which is entirely predictable as a dynamic relationship operating through natural selection where prey and predator exert different selection pressures on one–another. What is impossible to explain is just how this could, or would, be the work of an intelligent designer. Why design moths to be active at night to avoid being eaten, then redesign bats to find moths in the dark, then redesign moths to evade bats' newly–designed detection abilities, then redesign bats… well, I'm sure you get the picture. This could not conceivably be the work of a single intelligent designer because it is neither designed nor intelligent.

Garter Snakes and Rough–Skinned Newts.

The last example of a predator–prey arms race – and I'm afraid for those who don't like them it involves another snake, albeit a non–venomous species which is itself subjected to attack by a toxin. This is the arms race between the Common Garter snake and the Rough–skinned newt [20].

The Rough–skinned newt is a North American species which exudes a toxin from glands in its skin when attacked by a predator. Normally the acrid smell it also produces is enough to deter a predator but if the newt is eaten, it can be quickly fatal to the predator. The toxin is produces is tetradotoxin (TTX) which blocks the sodium pump mechanism which is an essential part of muscle and nerve function. Basically, when a muscle cell contracts or a nerve cell produces an electrical impulse, sodium ions flood into the cell and need to be pumped out again. This sodium pump relies on special tube–shaped proteins which cross the cell membrane and act as channels for the sodium ions. TTX binds to these proteins and prevents them working.

However, Garter snakes which inhabit the range of the Rough–skinned newt can prey on them, apparently without harm, whereas those Garter snakes from outside the newt's range, although still resistant, have a much reduced level of resistance.

It has been shown [21] that the Garter snake has a genetic mutation that slightly alters the protein in the sodium pump mechanism so TTX is hampered or prevented altogether from binding to it. This has created selection pressure on the newt to produce more and more potent TTX so it makes up in quantity what it lacks in quality so far as the Garter snakes are concerned. In areas where the two coexist, Rough–skinned newts produce many times more TTX than is needed to kill most potential predators.

In areas where they don't coexist, Garter snakes have a lower resistance, although thy still have some resistance, suggesting this may be an ancestral trait. Rough–skinned newts also have reduced toxicity. This suggests there is a cost to both species; the newts in producing more toxin and the snakes in reduced fitness with greater resistance. It has been shown that highly resistant snakes are in fact slower, making them more susceptible to predation themselves. For the Garter snake there is evolutionary trade–off between being slower and more susceptible to predation and having an additional food resource.

As well as illustrating an arms race, this illustrates that, in the absence of something like a food resource that can be exploited due to a mutation, the mutation may be detrimental; in the presence of that resource the mutation can have a net benefit. The environment in which the carrier of that mutant allele exists determines whether the mutation is beneficial or detrimental, so the environment determines the meaning of the changed genetic information. The meaning can change even if the amount of information remains the same. We will see another example of this when we look at the malaria parasite and human host relationship.

It should be obvious by now that this is yet another example of where a designer would need to be mad to design the Garter snake and the Rough–skinned newt in such a way that one kept on providing him or her – or it – with problems but the solutions to those problems were problems from the point of view of the other species. Can you imagine having an arms race with yourself where one half of you does not know what the other half is doing and where you invent even different solutions to the same problem?

In fact, where is the intelligence in creating living things that eat one another? Yet very many do with all the accompanying pain and suffering that often accompanies it. In the next chapter we will look at the relationships between parasites and hosts. As well as examples of arms races, these are examples of the sheer malevolence any designer would need to have to design such things.

The Unintelligent Designer

4. Parasites

"I often get letters, quite frequently, from people who say how they like the programs a lot, but I never give credit to the almighty power that created nature, to which I reply and say, "Well, it's funny that the people, when they say that this is evidence of the almighty, always quote beautiful things, they always quote orchids and hummingbirds and butterflies and roses."

But I always have to think too of a little boy sitting on the banks of a river in West Africa who has a worm boring through his eyeball, turning him blind before he's five years old, and I reply and say, "Well presumably the god you speak about created the worm as well," and now, I find that baffling to credit a merciful god with that action, and therefore it seems to me safer to show things that I know to be truth, truthful and factual, and allow people to make up their own minds about the moralities of this thing, or indeed the theology of this thing."

David Attenborough

The world of parasitology is a rich seam of embarrassment for creationists because so much of it seems to epitomise pure evil, if such a thing exists. It is also a rich seam of the evolutionary arms races which we saw in the previous chapter make such a mockery of the idea of intelligent design. With parasites you have arms races and what can only be described as uncaring, amoral, mindless selfishness (if selfishness can be described as mindless).

Leaving aside the question of intelligence and design, where is the compassion? Can this really be the intelligent design of a maximally good god?

And it is here that ID proponents often go off message and fall back on religious fundamentalism and start talking about 'Eve's sin' and 'The Fall' [22] which 'allowed evils to enter the world', conveniently forgetting both that they are supposed to be presenting an alternative science to evolution and that the Christian god actually claims credit for creating evil in Isaiah 45:7. ID then becomes nothing more than fundamentalist religion and one that cherry–picks from the holy books.

Before I look at specific examples, imagine a scenario in which a man in your neighbourhood is obsessed with getting revenge for some wrong or other that he feels was done to him many years ago, so he spends his time trying to invent new ways to randomly harm and even kill the people in his neighbourhood. The neighbours try to defend themselves against him but he cleverly finds ways around their defences and improves the poisons so their antidotes no longer work. The people have to adopt special measure to ensure he doesn't get through to them and live in constant fear of what he might do. They spend a huge proportion of their income on health care facilities just to provide treatment when he succeeds, but he even finds ways around their medical treatments.

But it is worse than that even. The man isn't content with making human lives a misery but he does the same to the whole of nature; plants, animals, just about everything. Noting is safe from his inventive genius as he constantly looks for ways to harm and kill them.

Then, in an even more bizarre twist, the man seems to forget his original intention – of making things sick – and now switches sides and tries to help minimise the harm he is doing by helping his victims overcome his attacks! He doesn't stop his attacks, but he tries to invent solutions to his own created problems, some of which are more effective than others but none of which are as effective as stopping the attacks would be.

Creationists want us to worship this psychopath, marvel at his genius and sing songs in praise of him. They teach their children to look to it for moral guidance as the person who defines right and wrong.

I would suggest that in any normal society this man would be kept in a high security facility and would be notorious as the epitome of evil; as a depraved monster who sadistically plays with lives and causes misery and suffering for the fun of it or for motives which are too bizarre and insane to comprehend. He would be someone for decent, compassionate and caring people to threaten their unruly children with.

I will illustrate this analogy with a few of the very many examples of what creationists want us to believe is the work of a maximally good, highly intelligent designer.

The Crypt–Keeper Wasp.

The Crypt Gall wasp is an innocuous wasp that parasitises the curiously–named Sand live oak tree. The female bores a hole in the tree's stems and lays an egg at the end of it. The presence of the egg induces the tree to produce a crypt gall in which the grub lives, eating the inside of the gall until it is ready to leave and pupate. It then eats a hole through to the outside world, escapes from the gall and pupates.

However, a few of them don't wait that long. Instead, they eat their way to the surface, stick their head into the hole to plug it and eventually die. They are the ones that have been parasitised by a parasitoid wasp, the Crypt–keeper wasp which has laid an egg in the body of the grub [23]. The presence of the egg and the grub it hatches into turns the Crypt Gall wasp grub into a zombie.

Having taken control of the grub and prepared its escape route, it sets about eating the grub from the inside until it is ready to leave, when it only need bore a small hole in the remains of the Gall wasp grub's head, neatly getting round the problem of not having jaws strong enough to bore through wood.

Creationists would have us believe that this is the intentional design of a maximally good designer. Seriously! Imagine a man in your neighbourhood who did things like that! Would you look to him for moral guidance and teach your children to admire him? Creationists will also tell you this is because of something two human did once, as though this has anything to do with humans or has any effect on us by way of punishment for something someone else did.

Dinocampus coccinellae and Ladybirds.

Dinocampus coccinellae is another pretty little member of the wasp family. Like a lot of members of this family, it is a parasitoid species, the females of which lay their eggs in the living bodies of other creatures, often other insects. This particular one lays its eggs in ladybirds.

For several weeks an infected ladybird will carry on hunting aphids as though nothing much is happening and oblivious of the maggot growing inside it and eating its internal organs, apart from those essential to the ladybird, obviously. The ladybird needs to be kept alive to feed the growing wasp larva but unlike many related species of parasitoid wasp, it doesn't kill its host when it emerges from its body because it needs it for one more thing. It uses it as a living shield, making use of the ladybird's warning colouration to deter would-be bird and animal predators, and the ladybird's twitching reflex to deter ants, etc.

It does this by turning the ladybird into an automaton which stands over the pupa until the adult wasp emerges and flies away to find a mate and repeat the cycle.

The mystery was just how this conversion to an automaton was precisely timed to coincide with the wasp larva emerging from its body until a team led by parasitologist, Nolwenn Dheilly, of the University of Perpignan in France, discovered the answer [24]. As well as an egg, the adult wasp injects the ladybird with a virus which it had been storing in its oviduct and which multiplies rapidly and infects the ladybird's brain, but appears not to do it any harm - yet. This virus is new to science. Meanwhile, the developing larva produces a chemical which suppresses the ladybird's immune system.

When the larva leaves the ladybird's body, immunosuppression is switched off, the ladybird's immune system kicks in and promptly attacks its own virus–infected brain, leaving only the twitch reflex and enough nervous activity to keep it alive. Amazingly, a quarter of infected ladybirds recover and may go on to be infected again.

Now, here is an example of an evolutionary process which can't possibly be seen as either the work of a benevolent, caring, kind, designer or something designed to serve the needs of humans. Ladybirds are a major predator on aphids which do enormous economic damage to crops as well as spreading plant viruses, so why would an anthropophilic designer design something to harm a potential ally against aphids? And why design them to eat aphids if it was going to stop them doing so? Whatever designed this certainly hadn't got the welfare of humans in mind.

Does this designer have something against ladybirds, maybe? Did they commit some ancestral sin, and so have to be eternally punished with exquisitely nasty suffering?

The only things that gain in this are the virus and the parasitoid wasp which have formed a mutually beneficial alliance so that neither can reproduce without the other. Did the designer favour viruses or parasitoid wasps, or both, maybe? Does this designer favour parasites?

For its sheer malevolence, this takes some beating. Whatever intelligence that could design such a thing could never conceivably be called good, or benign, or worthy of praise and worship. It might be worthy of fear and loathing, repugnance even, but surely never of adoration. There are some close contenders, mind you -the beautiful little Emerald cockroach wasp (*Ambulex compressa*), another charming little wasp, for example.

As an evolved process, the product of an unemotional, amoral, mindless, undirected and unplanned evolutionary process, of course, it becomes easily understandable, even elegant, though no less exquisitely nasty and, scaled up a thousand-fold, the thing of nightmare.

Emerald Cockroach Wasp and Cockroaches.

The Emerald cockroach wasp is an exquisitely beautiful, iridescent little parasitoid wasp, which hunts cockroaches on which to lay its eggs. Anyone ignorant of the facts of its life–cycle might be forgiven for thinking it is the work of an aesthetic designer who designs beautiful things for the pleasure of it.

The truth however tells a different story. When it finds a cockroach it delivers a sting to the cockroach's thoracic ganglion that temporarily paralyses its front legs. This reduces its mobility so that the wasp can deliver a second, very precise sting to its head ganglion in a location that paralyses its escape reflexes. This sting turns the cockroach into a compliant zombie that cooperates with its own demise. The first thing it does is groom itself very thoroughly to remove and fungal spores or bacteria.

The wasp needs the cooperation of the cockroach because it is too small to pull or carry it into the burrow it prepared earlier. For reasons which are not clearly understood yet, the wasp bites off the ends of the cockroach's antennae then uses one of them to pull the now–compliant cockroach into the burrow where it lays a single egg on its abdomen. The wasp then fills the entrance to the burrow with small pebbles to deter other predators. The cockroach makes no attempt to escape but simply sits there as though in suspended animation.

When the egg hatches the grub lives for a few days as an ectoparasite before burrowing into the body of the living cockroach where it becomes an endoparasite, eating the cockroach's internal organs in an order that keeps it alive for as long as possible. After about eight days it pupates inside the cockroach's body to emerge some time later as an adult.

Neat, eh? Just imagine the mentality of something that would design that as a means of producing Emerald cockroach wasps – and for what end? Just to produce more Emerald Cockroach wasps.

The next example of the zombification of the host by a parasite involves another member of the hymenoptera; in this case an ant, not a wasp. However, rather than the hymenopteran being the parasite, this time it is the victim. The parasite is a fungus.

Ophiocordyceps unilateralis and **Carpenter Ants.**
Ophiocordyceps unilateralis is a fungus that infects Carpenter ants and takes control of their brain. Instead of doing what Carpenter ants normally do – foraging for food on the forest floor – an infected ant climbs high into the forest canopy, looking for a leaf with just the right humidity. Having found one, it crawls around to the underside, finds the central leaf vein and locks its jaws onto it.

The fungus then kills the ant and grows a fruiting body out of its head. This produces spores which fall to the forest floor and infect new ant hosts.

Where the ants have locked their jaws onto the central vein it leaves a dumbbell–shaped scar. These scars have been found on 28 million year–old fossil leaves from Germany, so the fungus has been pulling this trick on ants for at least 28 million years.

I'll turn now to that other group of parasites, the parasitic worms, again a rich seam of nastiness that creationists assure us were intelligently designed by a loving and merciful, omnipotent and omniscient designer.

Lancet flukes, Snails, Ants and Cows.

This example is yet another example of an ant being turned into a sacrificial zombie but this time it involves a snail, not as the parasite but as an intermediate host. The parasite is the flatworm known to science as *Dicrocoelium dendriticum* or the Lancet fluke, one of a group of trematodes. It spends its adult life in the bile ducts of cows and sheep, and sometimes humans [25].

Like many parasites, this creature has lost many of the characteristics of free–living flatworms, being reduces to a mere breeding machine, even having lost its digestive system because its host provides it with pre–digested nutrients. This feature of parasite evolution is another embarrassment to creationists because one of their dogmas is that evolution is impossible because it always involves an increase in complexity, and of course their 'design' argument bizarrely tries to ascribe complexity of any sort to an intelligent design process so the last thing they want is reduced complexity being called evolution.

But that's not the main point of this example so far as the argument against intelligent design goes. This is an example of both the malevolence of any putative designer and its incompetence since if the notion were true, the problem is one of its own making and the solution is bizarre almost beyond belief.

The problem is that when the Lancet fluke wants to breed, it sheds its eggs into the digestive tract of its host so they are promptly deposited outside the host's body. The problem then becomes how to get them back inside the parasite's host because its preferred hosts are especially averse to eating their own faeces. The 'solution' is to have a snail eat the eggs along with the ruminants' faeces.

Parasites

The problem is cows and sheep don't eat snails either. And snails don't like baby Lancet flukes living inside them so they encapsulate them as small cysts and excrete them in their slime trails. Now the baby parasites are back in the open again, stuck in snail slime that sheep and cows don't eat either.

However, ants **do** eat snail slime probably to get the moisture in them, and they eat the cysts the snails excrete, but that still doesn't get the baby flukes inside their target hosts. This is where the true malevolence of a creative mad genius comes in (if you believe in intelligent design). One of the juvenile flukes in the cyst goes to the ant's nerve centre and takes control of it, changing its behaviour. Now, instead of retiring to the ant nest when it gets dark, the infected ant seeks a blade of grass and climbs to the top of it – waiting to be eaten by a cow or a sheep. Job done!

The problem of the parasite laying eggs into the digestive tract of its host has been solved in a truly bizarre and mindless, uncaring and amoral fashion – and all because of 'Eve's sin', apparently.

It's not as though an entirely different solution doesn't exist for an almost identical problem involving a group of fungi [26]. These fungi live on the faeces of ruminant animals such as cattle. To get into these faeces or cow pats the fungal spores pass through the gut of cattle unharmed and without harming the animal.

The problem is how to get their spores into the cattle in the first place. This is made worse by the fact that cattle avoid eating grass close to their faeces – a ring of repulsion exists. So, to get their spores over this ring of repulsion, the fungi have an explosive fruiting body or sporangium. When ripe the sporangium explodes with enough force to propel the fungal spores up to 3 metres (10 feet) at speeds of up to 90 km (56 miles) per hour. The explosive force comes from pressurized fluid which is squirted out with the spores, so making them stick to the grass – where they get eaten.

Now, it shouldn't be beyond the capabilities of an intelligent designer to use this or a similar method to get the Lancet fluke larvae back into the cattle host using a similar solution, should it? Apparently it is, but then this is just another example of the same putative designer re–inventing the metaphorical wheel and coming up with a totally new way to solve a problem it solved earlier and then, apparently, forgot about.

But this example is even more embarrassing for ID advocates!

Although the fungi do no harm to cattle, cattle **are** harmed by a parasitic worm no less, that rides piggyback on this dispersal mechanism [26]. The worm is a lungworm, a species of nematode (*Dictyocaulus viviparous*) which causes bronchitis in cattle and some other ruminants. This worm faces the same problem as the fungi (and the Lancet fluke). The solution the putative intelligent designer has designed is for the lungworm larvae to climb onto the fungal sporangia and wait to be projected over the ring of repulsion.

So, a solution for one worm and a totally different, far more complex, solution for another, to solve the same problem the putative designer designed in the first place. ID advocates regard this as intelligent.

Schistosomiasis.

Schistosomiasis is a debilitating disease caused by flatworms called schistosomes. According to the World Health Organisation in 2018:

> Symptoms of schistosomiasis are caused by the body's reaction to the worms' eggs.

> Intestinal schistosomiasis can result in abdominal pain, diarrhoea, and blood in the stool. Liver enlargement is common in advanced cases, and is frequently associated with an accumulation of fluid in the peritoneal cavity and hypertension of the abdominal blood vessels. In such cases there may also be enlargement of the spleen.

The classic sign of urogenital schistosomiasis is haematuria (blood in urine). Fibrosis of the bladder and ureter, and kidney damage are sometimes diagnosed in advanced cases. Bladder cancer is another possible complication in the later stages. In women, urogenital schistosomiasis may present with genital lesions, vaginal bleeding, pain during sexual intercourse, and nodules in the vulva. In men, urogenital schistosomiasis can induce pathology of the seminal vesicles, prostate, and other organs. This disease may also have other long-term irreversible consequences, including infertility.

The economic and health effects of schistosomiasis are considerable and the disease disables more than it kills. In children, schistosomiasis can cause anaemia, stunting and a reduced ability to learn, although the effects are usually reversible with treatment. Chronic schistosomiasis may affect people's ability to work and in some cases can result in death. The number of deaths due to schistosomiasis is difficult to estimate because of hidden pathologies such as liver and kidney failure, bladder cancer and ectopic pregnancies due to female genital schistosomiasis.

The death estimates due to schistosomiasis need to be re-assessed, as it varies between 24 067[1] and 200 000[2] globally per year. In 2000, WHO estimated the annual death rate at 200 000 globally. This should have decreased considerably due to the impact of a scale-up in large-scale preventive chemotherapy campaigns over the past decade [27].

---

[1] Global Health Estimates 2015: Deaths by Cause, Age, Sex, by Country and by Region, 2000-2015
http://www.who.int/healthinfo/global_burden_disease/estimates/en/index1.html
Geneva, World Health Organization; 2016.

[2] Prevention and control of schistosomiasis and soil-transmitted helminthiasis
http://apps.who.int/iris/bitstream/10665/42588/1/WHO_TRS_912.pdf?ua=1
Geneva, World Health Organization: 2002.

The Unintelligent Designer

Unlike the flatworm we met earlier, *Dicrocoelium dendriticum* or the Lancet fluke, this parasite has a relatively simple life cycle. The larva enters the human body by piercing the skin when the victim is bathing in infected water. The worms then take up residence in various tissues where the females lay eggs. Some of these are passed out in urine and faeces to contaminate water whilst some remain in blood vessels where they cause blockages and immune responses responsible for damage to the organs. The eggs hatch in water and are ingested by water snails where they mature into the water-borne larva to infect humans again.

What are we to make of the reason for schistosomes? What possible purpose do these parasites serve? In terms of design, what are they **for**, exactly? They appear to have no other function but to make people sick and cause stunting, anaemia and a reduced ability to learn in children.

They also seem to make sexual intercourse difficult and painful yet they are not sexually transmitted. Infection bears no relationship to sexual activity or sexuality so can't be explained by the traditional creationist excuse as punishment for sexual activity other than that approved of by religious fundamentalists.

If an intelligent designer really wanted to make people sick, why would it go to these lengths of creating a hugely complex organism then designing a delivery system that appears to select victims entirely at random? Surely, it would not be beyond the wit and power of an omnipotent deity simply to make people sick, if that was what it wanted to do. And yet we are expected by the ID lobby to believe that this is a highly intelligent way to achieve that objective, and of course a designer deity wanting to punish us for some sin for which none of us could conceivably be held responsible, is not at all fundamentalist theology; it is science!

We can look now at a few more examples of parasitic worms before we turn to micro–organisms such as bacteria and protozoa.

Liver Fluke (*Fasciola hepatica*).

The Liver fluke is a parasite on herbivores such as cattle and sheep and humans. It has a complex life cycle which begins when a female migrates from the liver of the host to the biliary duct where she can lay up to 25,000 eggs per day. Like the Lancet fluke, the eggs are discharged into the intestine of the host and are discharged in faeces. If they land in water they hatch into a larval form known as a miracidia. The miracidia seek out and infect a water snail.

In the snail, it goes through three stages, turning from a sporocyst to a redia and finally a large–tailed cercaria which exits the snail and finds aquatic vegetation in which it forms a cyst called metacercaria. If the vegetation is eaten by cattle, sheep or humans, the metacercaria, having been protected against the stomach acids by its hard case, emerges from the cyst in the duodenum and burrows through the intestinal wall into the peritoneal cavity. From there, it migrates to the liver and spends time eating liver cells until ready to breed and repeat the cycle.

Tapeworms.

There are several species of Tapeworm, four of which infect humans. These and other species also infect other hosts. The adults take up residence in the small intestine of their host where they are supplied with a constant stream of pre–digested nutrients which they absorb through their skin, so they have no digestive tract, having lost it during their evolution into obligate parasites. Some species can grow to several metres in length.

The head, or scolex, has an array of hooks with which the worm attaches itself to the wall of the intestine. The rest of the worm consists of a series of maturing segments, new ones being created at the head end as ones at the distal end are shed. Each segment acts like an independent, hermaphroditic organism which can produce fertile eggs asexually or can 'mate' with another segment of the same worm or that of another worm. When full of eggs the segment is released from the end of the worm and passes out of the host as a cyst which may be eaten by a secondary host – fish, cattle, pig, rodent.

When in the intestine of the secondary host, the eggs hatch and break out of the cyst. The larvae bore through the intestinal wall and migrate to the muscles of the secondary host where they form another cyst. If this gets eaten, without being killed by cooking, the cyst hatches (excysts) in the host intestine and the cycle begins again.

The only purpose of this elaborate and hugely complex system, including the complexity of the Tapeworm and its segments, appears to be to produce more Tapeworms – and to make the host sick. In humans, the symptoms of infection include: anal itching, bloody diarrhoea, diarrhoea, headache, increased appetite, insomnia, loss of appetite, muscle spasms, nausea, nervousness, seizures, stomach ache, vomiting, weakness and weight–loss. There are no known benefits apart from, in obese people, weight loss up to a point. The risk of other symptoms however more than offsets any benefits from weight–loss, which can be achieved in other ways with none of those risks.

So, the 'purpose' of this 'design' appears to be to make more Tapeworms, the better to make more hosts sick.

Incidentally, these examples show another aspect of biological 'design' – prolific waste. Tens of thousands, even millions, of eggs or larvae are created just to produce a few adults. Could an intelligent, omnipotent designer really not devise a less wasteful method of reproduction? Do creationists not insist that their creator god knows exactly which of the thousands of offspring will survive to adulthood? Why then not produce just that one? We will see more examples of prolific, wanton waste later on.

Giant roundworm (*Ascaris lumbricoides*).

The Giant roundworm is a nematode worm that infects some 25% of the human population worldwide. It is passed on when eggs are accidentally swallowed from dirty fingers or contaminated food or water. Larva hatch from the eggs, penetrate the intestinal walls and enter the circulatory system, via which they pass into the lungs, where they develop for two weeks before travelling up the respiratory tract to the throat, to be swallowed again as adult worms.

The worms attach themselves to the intestinal wall and get nutrients from the host. Adult female worms are 20-35 cm long and 3-6 mm in diameter. Males are 15-30 cm long and 2-4 mm in diameter. On mating a female can produce 200,000 eggs a day and can live for two years. The eggs are passed out of the host in faeces and can survive for several months until swallowed by a human to complete the life-cycle.

In Humans, symptoms include blockage of the biliary tract, diarrhoea, fever, nausea, obstruction of the bowel (which can be fatal), stomach ache, vomiting, weakness. In children symptoms can include slower growth–rate. During the larval stage, when the larvae are developing in the lungs, symptoms may include breathing difficulty, cough and/or coughing up blood and eosinophilic pneumonitis.

There are no known benefits to the host.

Purpose? To produce more Giant roundworms apparently or, if you follow the logic of the creationist explanation for the harm they do, to make their hosts sick because of 'sin'.

River Blindness (*Onchocerca volvulus*).

River blindness is the second most common cause of blindness world–wide; second only to trachoma, itself a triumph for the intelligent designer if you believe in such a thing.

It is caused by a parasitic worm, *Onchocerca volvulus,* spread by blood–sucking Black flies of the *Simulium* genus. The disease affects about 15.5 million people, 800,000 of whom suffer complete or partial loss of vision.

Infection enters the body when an infected Black fly takes a blood meal and injects the worm larvae beneath the skin. These migrate to the skin to form nodules in which they become adult male and female worms. They can live for up to fifteen years, protected from the immune system by the walls of the nodules they live it. Here they mate and the females produce up to a thousand microfilariae a day into the host's subcutaneous tissues, where they move freely around the body and enter the circulation.

These microfilariae are mostly responsible for harm done to the host. They can live for one to two years and when they die they give rise to many of the symptoms, including intense itching, loss of skin pigmentation and rashes. A great deal of this is caused by the surface proteins of a bacterium of the *Wolbachia* group that lives symbiotically within the cells of all stages of the life cycle of *O. volvulus* and on which it appears to be dependent [28].

Microfilariae also migrate to the eyeball where they cause inflammation and other complications that lead to blindness.

When a Black fly takes a blood meal from an infected person, microfilariae are transferred to the Black fly where they develop to form infective larvae, ready to be transferred to the next host [29].

So this system has everything, including the added complexity of a symbiotic bacterium on which the parasite depends. It has prolific waste, needlessly complexity – why can't an omnipotent creator just randomly make people go blind, if that's what it wants, and why go to these lengths if the plan is simply to make worms of the *O. volvulus* species? It even incorporates a design to overcome the host's immune response that, presumably, the same designer designed to overcome the parasitic worms, and other parasites.

Oh yes! It's all to do with the 'scientific fact' that a legendary couple once shares a stolen apple, of course! That makes good scientific sense.

Parasites

Enough with the parasitic worms! Let's turn now to the very many microorganisms – those single–celled bacteria and protozoa that live in and on us and just about every other living thing.

The Unintelligent Designer

5. Microorganisms

There are literally thousands of pathological, single–celled organisms that live on and in living things, making them sick and frequently killing them.

By no means all microorganisms are pathological of course; in fact the vast majority are harmless and many are beneficial. Some live symbiotically in the guts of other creatures and some live symbiotically inside cells. Some, as we saw with the *Wolbachia* group of bacteria in the previous chapter, are essential to many creatures although they are parasitic at least to a degree in many arthropods.

Whole books could be devoted to endosymbiotic bacteria but one that is of special interest to us is that on which the Tsetse fly is dependent for its ability to breed.

Tsetse Fly, *Wigglesworthia* and Trypanosomes.
The following is based in part on an article I wrote in 2013 [30].

Superficially, the Tsetse fly resembles most other typical flies but, due to some important anatomical difference it is placed in its own separate family the *Glossinidae*. All 33 species belong to the single genus *Glossina*. Fossil Tsetse flies have been found in 34 million year-old deposits in Colorado, USA, so we know the family is fairly ancient. In Africa, diseases carried and transmitted by Tsetse flies kill an estimated 250,000 - 500,000 people and 3 million cattle a year, and costs an estimated 1-1.2 billion US$ a year (2010 figures).

In an article on the history of the African trypanosomiasis, Dietmar Steverding, said:

> "The prehistory of African trypanosomiasis indicates that the disease may have been an important selective factor in the

evolution of hominids. Ancient history and medieval history reveal that African trypanosomiasis affected the lives of people living in sub-Saharan African at all times." [31]

Tsetse flies have an interesting and rather unusual life cycle. A single female breeds about four times a year and can produce about thirty broods in a lifetime. However, unlike most other insects, which produce many eggs at a time and so rely on just a small percentage reaching adulthood to maintain the population (and so, incidentally, producing a constant supply of larva as prey for species higher up the food chain) Tsetse flies have evolved a different strategy. They produce just a single egg at a time.

Normally, with most higher insects, the eggs are laid and hatch and the larvae grow, shedding their skins at stages, so having several 'instars', each instar being larger and sometimes distinctly marked. Female Tsetse flies however retain the single egg in a uterus where it hatches. The larva lives and feeds for the first two instars inside the uterus where it is fed on a milk-like substance (I'm not making this up!) produced by a special gland in the uterus, complete with teat (I'm really not making this up!). The Tsetse fly suckles its young internally and gives birth to a large offspring in the form of a third instar maggot. This process of egg retention, development and feeding from a gland in a uterus in insects is known as adenotrophic viviparity.

Third instar larvae of other higher insects tend to spend a while in this stage laying down their final stores of nutrients to see them through pupation and into adulthood, and sometime even through adulthood since some fully formed insects don't feed at all but simply exist to mate and lay eggs, then they die. The Tsetse fly third instar stage lasts only for a short period however, during which it finds some earth soft enough to burrow into, gets itself a few inches below the surface and sheds its skin for the last time to form a hard-cased pupa. This third instar stage lasts such a short time that it has rarely been observed in the wild. We know about it from observing laboratory strains.

What this process means is that the female Tsetse fly has to supply her single offspring with enough nutrition to grow an entire adult Tsetse fly because the larva doesn't feed at all outside her body but pupates and develops into an adult fly using only the nutrients she supplies to it. This represents a huge investment on the part of the female who not only has to find a mate but also enough food for itself and its developing offspring. It literally needs to eat for two.

It does this by feeding off the blood of mammals, which it finds with a sophisticated heat-seeking system.

This, as with very many other blood-sucking insects like mosquitoes, has opened the Tsetse fly up to exploitation by a parasite, in this case a group of protozoans known as trypanosomids. Typically, these parasites have at least two hosts, spending part of their life-cycle in each. In this case the hosts are Tsetse flies and their victims - humans and other mammals, including herds of domestic livestock.

Tsetse flies don't appear to be affected by the trypanosomes although it is possible their behaviour could be modified. Humans, however, suffer from the debilitating and fatal disease, trypanosomiasis or sleeping sickness. Their livestock suffer from a variety of diseases including nagana from the Zulu *N'gana* meaning powerless/useless. Most wild African animals are resistant to it, having evolved for millions of years in its presence. Domestic animals, mostly imported into Africa in the last few thousand years however have no such resistance.

This begs the question, then why are humans not resistant since they and their ancestors have lived in Africa for many tens of millions of years? The answer to this is that it is very likely, because of our narrow genome range, that humans went through an evolutionary bottleneck of maybe just a few thousand individuals possibly just an isolated group in one small area. We also know that some of our closest ancestors lived in the Afar region of Ethiopia, from which the Tsetse fly in absent, so we may all be descended from a population which never actually had much contact with the Tsetse fly until relatively recently. Unlike zebra, wildebeest, elephant, giraffe and rhinos, we are comparative newcomers to much of Africa.

The Tsetse fly, together with it protozoan parasite, probably had a profound impact on human development in sub-Saharan Africa where it prevented two things which were of major importance to the rest of Euro-Asia and North Africa:

- Establishing herds of cattle to supply meat and (later) milk and dairy produce like butter and cheese.

- Using the horse as a beast of burden and as a source of energy for work such as ploughing, harvesting and threshing, so manpower remained the sole source of energy. A man can just about, using only his own labour, feed himself and his family by agriculture, with maybe a small surplus for trade or barter if the soils is especially fertile and watered. Otherwise the only existence possible without a source of energy greater than manpower is that of hunter-gatherer. Anything more requires at least beasts of burden and draught like horses, donkeys or oxen

It is believed that the single most important cause of sub-Saharan Africa remaining economically and technologically underdeveloped was the Tsetse fly coupled to the fact that there are no domesticable wild African animals, unlike Euro-Asia from where almost all human domestic animals came originally. Imagine if Bantus had domesticated the rhinoceros both as a working animal and for war. How would the Roman legions have fared against a Bantu cavalry mounted on rhinos? We could well have seen seventeenth and eighteenth century West African slavers raiding Western Europe and up into the Mediterranean to supply slaves for the new African colonies in the Americas and in South Asia and the Pacific whilst their priests condemned the white-skinned races as sub-human, backward and barbaric, fit only to be toilers and beasts of burden for the superior black races.

What has all this got to do with creationism and 'Intelligent Design'? Just a couple more things to point out, then I'll get on to that and pose a few questions for those creationists who have managed to get this far. If they've seen the questions coming they'll most likely have scuttled off somewhere by now to save the embarrassment of avoiding them later.

The trypanosomes which Tsetse flies infect humans and their livestock with, is a flagellate protozoan, i.e. it has a flagellum. The flagellum is one of the leading 'Intelligent Design' exponents, Michael J. Behe's favourite example of what he claims to be 'irreducible complexity', a concept with which he made his name and his fortune and which he still pushes as scientific 'proof' that the flagellum could not have evolved and so must have been intelligently designed. This is despite the evolution of the flagellum being known at the time he wrote his book, *Darwin's Black Box* [32], and despite being forced to admit under cross examination in the Kitzmiller vs Dover Area School District trial that there was no peer-reviewed scientific support for intelligent design [33] – as quoted on the inside front cover of this book.

But okay, for the purpose of this book, let's let Michael Behe have his intelligently designed Tsetse fly trypanosome flagellum in the knowledge that it'll come back to bite him soon.

The other little snippet of information about the Tsetse fly is one I personally find fascinating, showing as it undoubtedly does, how genes acting selfishly can produce mutually beneficial cooperation, contradicting creationists claims and quite paradoxical to what one might expect from a superficial understanding of evolution and a misunderstanding, deliberate or otherwise of the term 'selfish gene'.

According to R.H. Gooding, of the Department of Biological Science, University of Alberta, Alberta, Canada, and E.S. Krafsur, of Department of Entomology, Iowa State University, Iowa, USA [34]:

> Each Tsetse species harbours from one to three prokaryotic symbiont species, and these symbionts may provide

opportunities to reduce the vector competence of Tsetse flies. The most important symbiont, *Wigglesworthia glossinidia* , resides in a special bacteriome in the anterior part of the midgut and probably was a symbiont in the ancestor of all extant Tsetse species. It probably produces one or more substances that are essential for Tsetse reproduction. *Sodalis glossinidius*, a secondary symbiont not known to be essential for any Tsetse species, is found in the midgut and other tissues of several Tsetse species. *Wolbachia* is found in the gonads of some Tsetse species and is probably inherited through a strictly maternal lineage. Its effects on Tsetse flies have not been established, although in other insects *Wolbachia* has a variety of effects on their hosts, including inducing cytoplasmic incompatibility.

So, Tsetse flies would not now exist if it were not for a group of bacteria with which they have formed a symbiotic alliance and which are essential for their reproduction. For that matter, neither would their co-symbiont *W. glossinidia*. This alliance evolved out of a host-parasite relationship because it was mutually beneficial to both species.

Okay, so let's look at this from the point of view of an intelligent design proponent [creationist quote mine alert!]:

Surely, the Tsetse fly is a wonderful example of intelligent design, isn't it? It has beautifully 'designed' heat-seekers system so it can find the highly nutritious food it needs to supply both itself and the young Tsetse fly growing inside it. It has wings which enable it to fly like an Exocet straight onto its target once the heat-seekers have locked on. It avoids the waste of laying lots of eggs and producing lots of offspring only as food for other species like so many other insects do, by protecting and nurturing its young in a specially designed uterus completes with milk-producing breast and teat. And, because most African wild mammals are immune to the parasites it carries, it can feed without risking killing its hosts - always a useful strategy for a parasite.

The only problem is that it is quite easy to explain how **all** of these systems could have evolved. We know that they have had enough time to evolve because we know that Tsetse flies have been around for at least 34 million years, so we can explain them all by a known, observable natural process without needing to add the infinite complexity of an unexplained supernatural entity whose own 'design' remains to be explained. The existence of this supernatural entity remains not even a hypothesis but a mere notion (and merely one notion amongst an array of such notions limited only by human imagination) so multiplying entities needlessly. Hence evolution is the most vicarious hypothesis and the only one which is scientific in that it is the only one which is theoretically falsifiable and the only one to contain verifiable entities.

Evolution also explains the otherwise inexplicable inclusion of an obligate symbiotic bacterium in the Tsetse fly reproductive system, including the special anatomical adaptations for housing them. There is simply no rational way this can be described as intelligent design. No amount of special pleading can make this design look like the work of an intelligence in any normally employed use of that term.

Why would an intelligent designer infect its creation with trypanosomes, complete with their 'irreducibly complex' flagella? Was it so they would have something nasty to give their hosts by way of a thank you for the meal they just took, or so their hosts would then need to be intelligently designed to resists it? Or was the intelligent designer really only interested in trypanosomes?

And lastly, did your intelligent designer **really** create the flagellum of the trypanosome and create the Tsetse fly as its vector, complete with the *Wigglesworthia glossinidia* bacteria to allow it to breed, just so it could kill 250,000 - 500,000 Africans a year together with 3 million of their cattle, prevent Africans from being able to benefit from domestication of cattle and beasts of burden and draught, and hold Africa back in the early iron age both economically and technologically and unable to exploit her natural resources and metals that most of the developed world was able to use, for most of its history?

If so, in what sense of the word 'love' was it the act of a loving god and how can we distinguish such a god from a malevolent, evil god, or a mindlessly unintelligent stupid god whose 'plan' is indistinguishable from no plan at all?

This is the hardest point for the intelligent design movement to explain, bearing in mind that they are almost invariably Bible literalists too. They deny any connection of course, and expect us to believe in the fantastic coincidence of them all just happening to be religious fundamentalist, or at least subscribers to the Abrahamic creation myth from the Book of Genesis, including the fundamental belief that the earth and all its creature were created by the 'intelligent designer' for mankind.

Fortunately, science has no such conundrum to cope with so we don't need the mental gymnastics and moral ambivalence creationists need to cope with these little obstacles reality keeps putting in their way. And we can marvel at the process that created the Tsetse fly without needing to dismiss it and without inventing barbaric, patronising, condescending, racist and judgemental reasons to explain what Africans did to deserve it.

And nor can we use that latter invention as an excuse not to help do something about it.

Sleeping sickness or trypanosomiasis, caused by trypanosomes carried by Tsetse flies, although economically and historically important for the development of Africa, is not the biggest killer. That 'honour' must go to malaria carried by mosquitoes.

Malaria and Mosquitoes.

If the putative intelligent designer has anything to be proud of, it should be malaria for the way it devastates communities and kills so many children. Before you take offence at that, remember, if you believe in a creator god you believe he knew exactly what his creation would do before he created it. He knew precisely the effect malaria would have on people and created for that purpose.

So highly regarded is the supposed designer of malaria that Michael J. Behe has written a book [35] citing the way it has designed a way round human medical science's attempt to control it with anti–malarial drugs as a wonderful example of the designer's inventiveness, claiming that it must have been intelligently designed. The book has been comprehensively refuted by Kenneth R. Miller, professor of biology at Brown University, Providence, Rhode Island, USA [36] but never–the–less Behe and his supporters continue to argue that it is evidence for intelligent design, even demanding Miller apologise for his devastating refutation.

Behe's book was a sequel to his *Darwin's Black Box* in which he credited the supposed intelligent designer with creating the flagellum of the bacterium, *Escherichia coli*, which in its pathological forms helps it make humans sick and die. I'll have more to say about this later when I deal with the 'irreducible complexity' argument, which seems to be about the only argument for ID that creationists can come up with.

Meanwhile, this section is about malaria.

The World Health Organization reported 216 million cases of malaria in 91 countries in 2016, an increase of 5 million over 2015, with 445,000 deaths [37]. 90 percent of the cases were in Africa, as were 91 percent of the deaths.

Malaria is caused by five species of *Plasmodium* parasites, of which two species are the main problem – *Plasmodium falciparum* and *P. vivax*. These are spread by female mosquitoes of the *Anopheles* genus of which there are some 400 species. Around 30 of these are malaria vectors. Clearly, the Intelligent Designer left nothing to chance and went for over–kill in its chosen delivery systems.

Female *Anopheles* mosquitoes need a blood meal to feed their developing eggs. To ensure a good flow of blood as they feed, they inject a small quantity of saliva which prevents blood–clotting and is responsible for the reaction to the bite. If the female is infected with the *Plasmodium* parasites, these will be injected along with saliva. Once in the host's circulation the parasites migrate to the liver where they reproduce asexually to produce thousands of merozoites. These migrate out of the liver into the blood and enter red blood cells where they again reproduce asexually to produce 8–24 new merozoites. At that point the infected blood cell bursts and the merozoites infect new red blood cells.

Some of the merozoites develop into immature gametocytes which are the precursors to male and female gametes. These are taken up by the female mosquito when she takes a blood meal and male and female gametocytes mature and fuse in her gut to produce a motile ookinete which develops into a sporozite which migrates to the mosquito's salivary glands, ready to be injected into the next host to repeat the cycle.

It's even more interesting than that, though, and something which is quite easy to understand in terms of evolution.

- In 2017 a team of researchers from *Institut de Recherche en Sciences de la Sante(IRSS), Bobo-Dioulasso*, Burkina Faso and *Maladies Infectieuses et Vecteurs: Ecologie, Genetique, Evolution et Controle (MIVEGEC)*, Pontpellier, France, showed that female Anopheles mosquitoes are 25 percent more likely to take a blood meal when infected with *P. falciparum*. In other words, the malaria parasite is manipulating its mosquito vector's behaviour.

- One of the major problems facing human attempts to control and eradicate malaria is the spread of forms of the parasite which have become resistant to the anti–malarial drugs used.

Not content with designing this malevolent disease and its delivery system, creationism's putative intelligent designer saw fit to improve its ability to find its target – its mammalian victims – and is now redesigning the parasite to overcome human medical science's attempts to protect humans from it.

Remember, in the ID model, this is all for a purpose, so what purpose can be discerned in this system? Certainly, it results in more *Plasmodium* parasites. It also results in hundreds of millions of people, many of whom will be children, being severely incapacitated and dying. Is this the purpose? Let's run through the theology – what creationists believe about their supposed intelligent designer which, as well as being super intelligent is also perfect, all–knowing and all–powerful – the so–called 'omnis', omniscience and omnipotence and of course omni–benevolence.

In a debate about the zika virus, I once asked a creationist to tell me where I had gone wrong in my thinking when I gave him the following list of what he purported to believe (modified here to apply to *P. falciparum*, the major malaria parasite in terms of virulence and mortality). I asked him to answer yes or no to the questions – 'no' indicating that I had made a mistake:

1. Do you believe in an intelligent designer?

2. Do you believe this intelligent designer is benevolent and loves its creation?

3. Do you believe something designed by this intelligent designer would work as intended?

4. Do you believe this intelligent designer designed the malaria parasite, *P. falciparum*?

5. Do you believe *P. falciparum* causes malaria?

6. Do you believe this intelligent designer knew *P. falciparum* would cause malaria?

7. Do you believe malaria is a good thing?

8. Do you believe an intelligent designer who designed a parasite to cause malaria would be a benevolent designer who loves its human creation?

9. Do you believe deliberately creating malaria is morally wrong?

10. Do you believe this intelligent designer is morally good?

He never answered the questions and broke off the conversation. It is better for a creationist to avoid awkward questions than to confront their answers. Somehow, creationists manage to overlook the sheer nastiness in nature and see only the wonder and beauty, or they dismiss it all as the fault of human 'sin' and not something their supposedly omnipotent creator has any control over or responsibility for – or maybe deliberately causes for some ultimate, unknown and unknowable 'good'.

I think it would be quite difficult to explain to a person dying of malaria, or to the mother of a recently–deceased child, that it is all for some long–term good, or to satiate the need for revenge of some deity who really would like to be all–loving – if only he could get over someone eating his apple.

Trichomonas vaginalis.

Trichomonas vaginalis is a flagellated protozoan and a sexually transmitted parasite in humans. It also has some interesting features from and evolutionary perspective (and of course from an ID perspective).

Infection is usually asymptomatic in men, who can then inadvertently pass it on to a partner during intercourse. In women it causes trichomoniasis although some women can also be asymptomatic. It is the most common pathogenic protozoan infection in industrialized countries, there being some 160 million new cases annually world–wide with some 5–8 million new cases annually in the USA.

Complications of trichomoniasis include preterm delivery, low birth weight and increased perinatal mortality. It also predisposes to HIV infection and cervical cancer [38]. Infection can enter the urinary tract, fallopian tubes and pelvic cavity and can cause pneumonia, bronchitis and oral lesions.

T. vaginalis reproduces entirely asexually so never undergoes meiosis – the form of cell division that produces sexual gametes. However, researchers have shown that 27 of the 29 genes needed for meiosis are present in the genome [39]. Perhaps an ID advocate could explain these atavistic genes – complexity for the sake of it, maybe? Or are they simply there because an ancestor species once had sexual reproduction in its life–cycle and there has been no evolutionary pressure to remove them?

Which is the more vicarious explanation?

In 2007 a huge international team succeeded in sequencing the genome of *T. vaginalis* [40]. They found at least 26,000 confirmed genes (about the same number as the human genome) with a possible further 34,000 genes giving a massive total of about 60,000 genes. This very high number is probably accounted for by repeated gene duplication. Needless complexity?

The Unintelligent Designer

Lastly, there are signs of emerging resistance to the treatment of choice – metronidazole. There are two strains of *T. vaginalis* world–wide, one of which appears to carry a virus, the *T. vaginalis* virus (TVV). The presence of this virus appears to affect resistance to metronidazole! It also appears to make the pathogen more virulent.

Lastly, in a nasty little twist, when *T. vaginalis* is attacked and killed by metronidazole it releases the virus which provokes a reaction in the human cells, making the symptoms worse and posing an increased risk to mother and baby during pregnancy [41].

Has the Intelligent Designer really inserted a virus into *T. vaginalis* to make it more harmful, to help it overcome human medical science and to enable it to hit back?

At this point, I probably don't need to pose the question of benevolence. The ID answer would be the same any way – the 'scientific' explanation based on a literal interpretation of a book of origin myths from the 'fearful infancy of our species' (with thanks to Christopher Hitchens). And of course it has the 'irreducibly complex' flagellum, the better to perform its 'purpose'.

Entamoeba histolytica.

Entamoeba histolytica is a parasitic amoeba that predominantly infects humans and other primates. It normally inhabits the large intestine in its motile, trophozoite form, attaching itself to the intestinal epithelium where it feeds on bacteria and food particles. It digests bacteria with lytic enzymes, i.e., enzymes that destroy cells. If it remains in that form, the host will be asymptomatic.

Occasionally however, it uses these cell–destroying enzymes to attack the wall of the large intestine itself, inducing an immune response. The host's immune system responds by mobilising phagocytes which would normally engulf the microorganisms and destroy them with their own lytic enzymes.

Here, though the amoeba has a little trick up its sleeve; it kills the immune cells, releasing their lytic enzymes so these enzymes also attack the host cells. The amoeba then eats the dead cells.

Having bored through the intestinal wall, the *E. histolytica* trophozoites can enter the portal system blood vessels and be transported to the liver where a similar round of cell destruction takes place leading to amoebic liver abscesses. These abscesses can burst or the trophozoites can enter other tissues and organs where similar pathology occurs.

The trophozoites that remain in the large intestine continue to reproduce asexually and eventually form cysts or 'eggs' which pass out of the host in faeces. The cysts can survive outside the host in soil and water and on food. Their spread is aided by poor hygiene. They re–enter the host when contaminated food or fingers are placed in the mouth. Once in the intestinal tract they excyst and become active trophozoites again.

So here we have a nasty little pathogen that turns the host's immune response on itself – an immune response that presumably creationists believe their intelligent designer designed to protect us from the pathogens it designed to harm us.

And this, presumably, is its purpose – to make more potentially pathological amoebae to make us and other primates sick.

Could there be a better example of the mindlessness and lack of planning of biology or of the malevolence of this incompetent designer?

Probably, as we shall see.

Escherichia coli.

I will look now at that favourite of the ID movement, *Escherichia coli* (*E. coli*); a favourite of the ID movement because it featured in Michael J. Behe's book, *Darwin's Black Box* [32], which popularised the notion of 'irreducible complexity' as evidence for intelligent design, a book eulogised by the evangelical Christian right and pilloried by science generally and biologists in particular. Those who know nothing of the subject find it convincing; those who understand the subject point to the flaws in Behe's reasoning, and the errors in his science that remain uncorrected in subsequent editions.

In this book, Behe argued that the flagellum of *E. coli* is an example of irreducible complexity and so something that could not have evolved gradually. He argued that, since all its parts need to be present for it to work, and so provide something advantageous on which natural selection can operate, there is no mechanism for it to evolve gradually by an accumulative process such as Darwin proposed.

Before dealing with Behe's argument, which has been comprehensively refuted by Kenneth R. Miller, amongst others, I'll deal first with the bacterium and what its pathological forms do.

E. coli is the name of a highly diverse group of bacteria which displays far more genetic diversity than entire families of multicellular organisms with only about 20 percent of genes being common to all strains. Some authorities believe the taxon is overdue for revision and that some bacteria classified as *Shigella* should be reclassified as strains of *E. coli* while some *E. coli*, for example, the K-12 strain, should occupy a different taxon. Some strains have no flagellum, so lack motility.

E. coli is a normal part of the lower gut flora of warm–blooded organisms and may even be symbiotic in that it produces vitamin K_2 and helps prevent the growth of pathogenic organisms. It can live and thrive in anaerobic, i.e., oxygen deficient, conditions but can also live aerobically in fresh faeces where it can increase rapidly for a few days in the presence of oxygen. In normal conditions in the gut it can replicate every 20 minutes. Replication is normally asexual but they can exchange genetic material by conjugation.

It is parasitised by bacteriophage viruses which can incorporate *E. coli* genes into their own genome then pass these on to other *E. coli*, so also facilitating horizontal gene transfer. It is believed that strain 0157:H7 acquired the Shiga toxin by horizontal gene transfer from a *Shigella* bacterium.

The *Escherichia* bacteria are related to *Salmonella*, the two groups having diverged about 100 million years ago as the mammals diverged from the reptiles and birds. *Escherichia* species live in the gut of mammals while *Salmonella* occupies the same niche in birds and reptiles.

Some strains of *E. coli* can become pathological, causing diarrhoea that is normally self-limiting in healthy adults but can be fatal to children, especially in the third–world. The strain 0157:H7, with its Shiga toxin can be fatal to the elderly and very young, and to people with reduced immunity.

Normally harmless strains of *E. coli* can become pathological when out of their normal gut environment. For example, if they enter the urinary tract they can cause urinary tract infections.

In 2011, the enterohemorrhagic *E. coli* strain O104:H4, believed to have been carried on Egyptian fenugreek seeds, caused a major outbreak starting in Germany and spreading to 15 other countries. In 1996, 20 people in Wishaw, Scotland, died as a result of food poisoning caused by *E. coli* 0157, traced to contaminated meat from a local butcher.

To return for a moment to the replication rate of *E.coli* the reader may recall that in normal conditions, *E. coli* can replicate every 20 minutes. One of the trick creationists play is to take a sequence of steps that in themselves are highly unlikely, but which need to happen for a particular thing – a biochemical pathway, drug resistance, etc., to evolve, then they calculate the probability of all of those unlikely things occurring **in the same cell as a single event** and conclude that the probability is indistinguishable from zero. Of course, this conveniently ignores the fact that the mutations don't all need to occur together in the same cell but can occur at different time in the population where they can accumulate until they are present eventually in the same cell.

The maths is actually quite simple, so try this little mind experiment:

Given a stable population of a billion *E. coli*, replicating every 20 minutes, how many billion–to–one chance mutations will occur on average in a typical day? The answer is 72. In a typical year that would be 26,280 billion–to–one chances happening in that population.

Now it gets a little more complicated. If a billion–to–one mutation gave a just 1 percent increase in the carrier's ability to replicate, how long would it take until more than half of the population carried the mutation?

Because this organism is such a totem for the ID movement because of Michael J. Behe's assertion of irreducible complexity and therefore non–evolvability, I will now turn to that claim and its refutation.

There is nothing especially unique about the *E. coli* flagellum *per se* of course because flagella and cilia are present in very may microorganisms as well as specialised cells in multicellular organisms – sperm, for example.

Very briefly, a flagellum consists of the hair–like protein filament which projects from the cell surface and extends across the cell membrane into the cell where it runs between an arrangement of proteins which together make up a small rotary motor which spins the filament – the so–called proton motor. The power for the motor is provided by the usual cell power source, adenosine triphosphate (ATP). This is perhaps the nearest thing in nature to a wheel. Take any part of that assembly away and the flagellum doesn't work.

This is also true for many structures and whole organisms. Take a hand away and the arm ceases to function for grasping and fine manipulation; take the mitochondria out of a cell and the cell ceases to function. So, the fact that it renders it non–functional if part is taken away does not, in itself, establish that it could not have evolved.

Behe's argument is that it never could have worked during any hypothetical slow, one–step–at–a–time, evolutionary process since there was never any benefit until the whole thing was in place. His argument is a slightly more sophisticated form of William Paley's Watchmaker argument; the argument from design or teleological argument.

Behe's objectives here are those of the fundamentalist Christian *Discovery Institute* [42], the organisation set up specifically to undermine public confidence in science. These are to present Darwinian evolution as a theory in crisis and to present disguised Bible literalism as a serious (and better) alternative scientific explanation.

I deal with the teleological argument in Chapter 2 so I will just note here that these types of argument are nothing more than arguments from ignorant incredulity combined with a presuppositional false dichotomy and a non–sequitur; the fallacy that if we don't know how something happened naturally, it is not only unknowable but must therefore have been done by a presupposed supernatural agent for which no evidence has ever been found. In reality, not knowing how something was done does not give us licence to invent our own solution and, with no evidence whatsoever, proclaim it as the cause. That is both illogical and intellectually dishonest.

Like almost all creationist/ID arguments, there is never seen any need to establish *a priori* that this presupposed intelligent designer actually exists. What the argument depends on is a cultural assumption that this creative deity actually exists. Hence it is not an argument **for** the existence of the designer but spurious confirmation of pre–existing bias. The ability to designate an imaginary entity as the cause of an unexplained phenomenon does nothing to establish either the existence of that entity or the validity of the conclusion.

For example, I could claim to have closed every gap in scientific knowledge by ascribing cause to an invisible hippopotamus that lives in my loft, or to fairies at the bottom of my garden. This might raise questions about my sanity but it would do absolutely nothing to prove invisible loft hippos or fairies at the bottom of my garden actually exist outside my imagination. And it could never be passed off as good science.

But, is Behe right that Darwinian (or more accurately, neo–Darwinian evolution) can't account for these examples of 'irreducible complexity'?

The short answer is; no, he isn't.

What he failed to take into account was the power of gene duplication and the ability of natural selection to adapt these copies for new purposes. He also failed to recognise that there are thousands of different versions of the bacterial flagellum, several of which contain fewer components than that of E. coli. It cannot be true therefore that the E.coli flagellum could not exist in a reduced form.

The last thing Behe seems to overlook is the ability of natural selection to exapt redundant structures for new purposes.

The E. coli flagellum consists of 40 different proteins, but only 23 of these are common to all bacterial flagella. So, either the 'intelligent designer' created thousands of different flagella, solving the same design problem, and the same need, in thousands of different ways, or it is possible to change the basic design in thousands of different ways, and not breaks it.

To make matter worse for the irreducibly complex argument, it turns out that only two of these 23 proteins are unique to flagella. All the others are closely similar to proteins that have other functions in the cell. Gene duplication and mutation are quite capable of explaining this difference.

Microorganisms

It is also possible that the basic assembly was an adaptation of some other structure which served some other unrelated purpose, the loss of which was more than compensated for by motility. One such function might have been to extrude a protein filament, possible to attack a prey or to attach the cell to a surface. All that would have then been needed would be a small change to convert the extrusion mechanism into a rotation mechanism and the cell would have been motile, no matter how inefficiently. Natural selection would have refined and improved the mechanism over time, one small Darwinian step at a time.

Irreducible complexity is thus not evidence of non–evolvability.

Two analogies serve to explain how irreducible complexity can be produced without simultaneous assembly of all parts in a single process. The first is the mouse trap which Behe himself cites as an example of an assembly that needs all its parts to work. It is quite possible with a little imagination to think of an alternative use for every one of those components, so none of them would need to be specially invented or designed for the purpose. They did not need to be functioning as a mousetrap or anything related to trapping mice.

The second analogy is that of the stone or brick archway. No–one would think that a stone mason or bricklayer would try to hold all the components of the arch in place while putting the key stone in. What is needed is some scaffolding and a pre–formed support on which the arch is constructed until the whole is assembled, when the support and scaffolding is removed, leaving no evidence of them ever having existed.

The last microorganism I will use in these examples is that well–known cause of intractable hospital infections, *Staphylococcus aureus*, often abbreviated to staph.

Staphylococcus aureus.

Staphylococcus aureus is a normal skin bacteria, frequently found as well in the nose, respiratory tract and in women the lower reproductive tract. It is normally benign but can be a common cause of skin infections, entering through cuts and scratches or in hair follicles and sweat pores where it can cause boils, carbuncles, cellulitis and impetigo. It can also cause respiratory tract infections, sinusitis and food poisoning.

It is a serious contributor to wound infections following surgery and can cause pneumonia, meningitis, osteomyelitis, endocarditis, toxic shock syndrome, bacteraemia, and sepsis. In the USA, an estimated 50,000 people a year die as a result of *S. aureus* infections.

In its pathogenic forms it shows an interesting examples of earlier arms races. The body's normal response to infection is to mobilise white blood cells but *S. aureus* can produce protein toxins which attack the white cells and proteins that bind and inactivate antibodies.

S. aureus is notorious for having developed strains resistant to antibiotics such as methicillin which used to be the antibiotic of choice to treat *S. aureus* infections. The resistant strain, methicillin resistant *S. aureus* (MRSA) is now a major problem in hospitals and nursing homes.

Accepting the ID model for the sake of argument, we would have to conclude that the intelligent designer is actively working to ensure its creation is still able to make people sick despite the best efforts of our immune system (which it also designed for the purpose of defending us, presumably) and the best efforts of medical science to defend us against it.

This and the preceding chapter dealt with parasites of one form or another – there are hundreds more but they should suffice to illustrate that all is not wonderful or benevolent in nature, contrary to the starry–

eyed, blinkered and carefully filtered view of most creationists. In the next chapter, I will look at needless complexity.

The Unintelligent Designer

6. Needless Complexity

When discussing biological systems, the question of minimal necessary complexity all depends on purpose and this is where claims of design come up against another hurdle. Recall that in Chapter 1 I concluded that needless complexity is poor design, not good design. It is not evidence of perfect design but of imperfect design; not of intelligent design but of stupid design. A hallmark of good design is that it is minimally complex to achieve a given purpose or to provide a given function.

So, the first question is, what exactly **is** the purpose of any given organism? This question arises for ID advocates because purpose is an assumption of the design claim. If you are going to postulate a designer, let alone a supremely intelligent one, you have to relate that to some assumed purpose, otherwise there is nothing to measure the effectiveness or the intelligence of the designer against.

Surprisingly, although theists often claim their faith gives them a purpose, actual answer to the question "What purpose?" are few and far between and are normally such nebulous purposes as worshipping God/Allah or purposes like living a good life – a purpose that anyone can give to their life regardless of any religious belief. In fact, this is a theological/philosophical question, not a scientific or biological one and yet purpose is integral, indeed central, to the ID argument which purports to be science, not philosophy and certainly not religion.

There is simply no way you can look at any given organism or ecosystem and discern some ultimate purpose to it all. It's true to say that metabolic processes and pathways have a function in terms of keeping the organism alive and able to function normally, so you could perhaps examine the design of such a process and ask is this is minimally complex, but not for the ultimate function of the organism that depends on it.

The Unintelligent Designer

Certainly, you can take a highly anthropocentric view of the world and claim, as indeed the Bible does, that it was all created for mankind – that all the animals were created as "an help meet" for Adam and his descendants and all the plants as food for humans and their animals. But again, that is a religious, and a specific family of religion's, point of view, not a scientific view. If the ID movement depends on these religious views as part of its design claim, then it gives up any claim to be science.

Nevertheless, if we allow creationists their religious purpose for the sake of argument, how well do their intelligent designer's designs measure up as minimally complex designs?

As creationists continually point out, living things are fantastically complex in terms of the number of component parts and multicellular systems are even more complex than single–celled organisms, failing to understand that complex design is bad design if the complexity is unnecessary. Listening to creationists you get the impression they think complexity is somehow a good thing in itself; that the more complex a design is, the better it is.

The following is a random sample of three Christian and three Islamic unsolicited statements of their life's purpose gleaned from Twitter. They may not be representative but they certainly reflect what people of these religions have been taught to believe their purpose is, and they **are** supported by their respective holy books.

- Without God, my life would be meaningless. So of course I dedicate my life to Him!

- Without God, my life would be meaningless, purposeless.

- ...God, my life, literally would be meaningless without You. Thank you. I will sing praises to You for the rest of eternity...

- As Salaamu Alaykum to all my brothers and sisters! Live life with wisdom, avoid the strife, submit to ALLAH cause [sic] thats [sic] the purpose of life.

- Purpose of this life is none but to worship ALLAH (S.W.T) on the way of prophet Muhammad (S.A.W).

- On the contrary, faith in Allah gives the believer the purpose of life that he [sic] needs. In Islam, the purpose of life is to worship Allah.

What then do the holy books say about the purpose of life?

The hopelessly muddled story of Genesis even seems to have got the authors confused. Firstly, our purpose is to have dominion over the (previously created) animals:

> And God said, Let us make man in our image, after our likeness: and let them have dominion over the fish of the sea, and over the fowl of the air, and over the cattle, and over all the earth, and over every creeping thing that creepeth upon the earth. (Genesis 1:26).

Followed soon by giving the animals a purpose what with them now being created after mankind merely to provide Adam with 'an help meet' in an apparently revised version of creation, or maybe another go at getting it right.

> And the Lord God said, It is not good that the man should be alone; I will make him an help meet for him. And out of the ground the Lord God formed every beast of the field, and every fowl of the air; and brought them unto Adam to see what he would call them: and whatsoever Adam called every living creature, that was the name thereof. (Genesis 2:18-19).

Then there is an attempt to define a purpose for woman (not clear if this is in addition to being Adam's 'help meet', for which she was originally created as the animals God had created earlier didn't measure up, or instead of. It amounts to much the same thing though):

> Unto the woman he said, I will greatly multiply thy sorrow and thy conception; in sorrow thou shalt bring forth children; and thy desire shall be to thy husband, and he shall rule over thee. (Genesis 3:16).

And finally there is another go at giving men a purpose, which, by great good fortune, just happened to be working in the fields of those who employed people to write the origin myths.

> And unto Adam he said, Because thou hast hearkened unto the voice of thy wife, and hast eaten of the tree, of which I commanded thee, saying, Thou shalt not eat of it: cursed is the ground for thy sake; in sorrow shalt thou eat of it all the days of thy life;

> Thorns also and thistles shall it bring forth to thee; and thou shalt eat the herb of the field; In the sweat of thy face shalt thou eat bread, till thou return unto the ground; for out of it wast thou taken: for dust thou art, and unto dust shalt thou return. (Genesis 3:17-19).

And that seems to have been about it until we get to Ecclesiastes. Having originally said it makes no difference what we do in life:

> All things come alike to all: there is one event to the righteous, and to the wicked; to the good and to the clean, and to the unclean; to him that sacrificeth, and to him that sacrificeth not: as is the good, so is the sinner; and he that sweareth, as he that feareth an oath.

> This is an evil among all things that are done under the sun, that there is one event unto all: yea, also the heart of the sons of men is full of evil, and madness is in their heart while they live, and

after that they go to the dead. For to him that is joined to all the living there is hope: for a living dog is better than a dead lion. For the living know that they shall die: but the dead know not any thing, neither have they any more a reward; for the memory of them is forgotten. (Ecclesiastes 9:2-5)

The author then seems to have decided we do have a purpose after all. Alas, he only managed to come up with:

Let us hear the conclusion of the whole matter: Fear God, and keep his commandments: for this is the whole duty of man. (Ecclesiastes 12:13).

It's not clear which set of 'Commandments' are being referred to here, but the chances are, on past form, that they will not be a million miles removed from the ten listed in Exodus 34:17-26 which are almost all about observing rituals and giving stuff to priests. (Can you guess who wrote them?)

And this 'obey all the rules' rule is pretty much what Mohammad seems to have latched onto for the Qur'an, presumably not being able to come up with anything better and which still kept the clerics in charge:

And I did not create the Jinn and mankind except to worship Me... (Qur'an 51:56-58).

Presumably, this 'obey all the rules and worship God' rule is for the promise of a reward later – after you're dead and can't come back to complain.

And that really appears to be it so far as a God–given purpose can be discerned from the holy books of the Abrahamic religions. Other religions have purposes based on ideas or reincarnation and what you are reincarnated as depending on how well you behaved in life. In other words, the purpose of life seems to be to obey all the rules again in the hope of something better in the next life.

But why does any of that require a fantastically complex organism, and what of the animals and plants that are not any use to humans and are not even harmful so not even fulfilling a creator god's need for revenge? What exactly are **they** for? What purpose does their design serve?   Take. For example, the little–known slug mites.

The Slug Mites.

The Slug mites (*Riccardoella limacum and R. oudemansi*) is one of my favourite pointless creatures because it really does seem to serve no purpose at all other than producing more slug mites.

If you look closely at the body of large slugs and snails, preferably with a hand lens, you may be able to see tiny white creatures moving about on the surface of their skin and even moving freely in and out of the pneumostome or breathing hole. These are the Slug mites.

They eat the slime that all slugs and snails freely produce to keep their surface moist. There are some hints that they may sometimes become parasitic by burrowing into their host's skin and they may even suck blood, but they appear to do no harm at all to the slugs and snails they live on. The amount of slime they consume can't possibly present their host with a significant cost, so they can't even be said to control the slug and snail population.

Nothing eats them except another mite (*Hyoaspis miles*), so maybe their purpose it to provide food for this mite so that it too can make more mites? And that's about it – an amazingly complex, almost microscopic creature that seems to serve no discernible purpose other than maybe being eaten by another mite. Even though, as a mite which has become small and greatly reduced compare to other mites, so showing how evolution can lead to reduced complexity, it is still a hugely complex, multicellular organism with all the complexity of a single eukaryote cell multiplied many times over and differentiated into specialised cells and organs.

Intelligent design? Huge complexity for little or no discernible purpose other than a piece of food for something else! Would it really be beyond the wit of an omnipotent designer to design something much simpler for *Hyoaspis miles* to eat – slug and snail slime, for instance – even if it did feel some compulsion to design pointless mites? Needless complexity is a hallmark of bad, stupid design or no design at all.

This next example of a mite is, if anything, even more pointless. It appears to do nothing at all other than produce more mites. It is the eyelash mite.

The Eyelash Mites.

The Eyelash mite, *Demodex folliculorum*, lives in human eyelash follicles. A closely related species, *D. brevis*, lives in the sebaceous glands associated with hair follicles. In all there are about 65 different species of the genus *Demodex* living in the hair follicles and sebaceous glands of different mammals, for example, *D. canis* which lives in the hair follicles of dogs where it is regarded as commensal, i.e., simply living on but not doing the host any harm, therefore not regarded as a parasite.

In humans, about one third of children and young adults have them, about half of adults and two thirds of the elderly, but some studies have indicated these figures may be an underestimate [43]. They are normally harmless but they are regarded as parasites because they live on skin cells as well as sebum. However, they only rarely give rise to any symptoms. They are so small that they are invisible to the naked eye. Up to twenty–five mites can live in a single follicle.

The mites live for a few weeks during which they mate and lay eggs. The eggs hatch into six–legged larvae in three to four days which become eight–legged adults in about a week.

And that appears to be it. They have no predators and they only rarely cause a problem for their human host; so if that was their design purpose, the designer was singularly inept. Or is their purpose simply to produce more eyelash mites? It has been suggested that they may be in a symbiotic relationship with their host because they keep the hair follicle clean, but is a highly complex, multicellular mite the simplest way a designer could have designed for keeping a hair follicle clean – the hair follicle it designed in the first place?

We can, or course, take any organism and ask the same question of ultimate purpose of it.

I was asked recently by a fiend what is the purpose of butterflies and I was rally hard–pressed to think of one. They may be useful as pollinators but could an omnipotent, intelligent designer really not come up with a simpler means to bring male and female plant gametes together without designing butterflies, bees and other pollinating insects?

If we look at the actual design of butterflies, and especially their wing patterns, we see lots of examples of our old friend, the evolutionary arms race. The purpose of those 'designs' is simply to preserve the butterfly to allow it time to breed and produce the next generation of butterflies.

I will illustrate this with a couple of examples:

The Peacock Butterfly.

As a child I remember being enthralled when I came home from school to find the pupa in the large jar I had had for a few weeks on our front porch shelf, had turned into the most beautiful Peacock butterfly. Such beauty that had emerged from a greenish–grey, spiky chrysalis which had itself been the result of a rather ugly, spiny black caterpillar I had found on stinging nettles. I took it home and kept it fed on fresh nettles, curious to see what on earth this thing was going to turn into.

So what **is** the purpose of the eye–markings on a Peacock butterfly? I tried to explain this is an illustrated article a few years ago [44]. The first thing is to understand the appearance of the butterfly from the point of view of its predators and its potential mate. It almost certainly doesn't look pretty for humans to enjoy.

To attract a mate, the peacock butterfly needs to flash its markings otherwise it is fairly well disguised as a dead leaf. The problem is, that flash advertises its presence to a potential predator such as a bird.

However, to a potential predatory bird, those eyes look just like the eyes of a stoat or a weasel and this is accentuated by the eyes being on both wings and the little white marks which give the illusion of movement. To a bird, a flashing peacock butterfly looks like a pouncing weasel or stoat, especially when seen from the butterfly's head end.

It is very easy to understand how this system evolved a small step at a time with each little improvement being more effective. What doesn't make sense is why an intelligent designer would design birds to eat butterflies then redesign butterflies to frighten birds. And this is just a single designer, remember. Solving problems it has just designed as solutions to problem it designed earlier, cannot conceivably be describes as intelligent.

Let's consider the supposed design processes here:

Predatory mammals are designed to catch and eat birds. Birds, in turn are designed to avoid being eaten (the 'solution' of how to feed the mammals is now a problem to be solved for birds). The solution to that pre–designed problem is the startle reflex that makes birds take flight when they see, no matter how poorly, approaching mammalian eyes.

The next 'problem' is how to design a solution to the problem of birds being designed earlier to solve the problem of how to feed birds. The solution is to use the bird's startle reflex against it! A designed solution to the problem of birds being eaten is now a problem for the bird in that it finds food harder to catch!

And this is regarded as highly intelligent!

So which is the more vicarious explanation? A mindless, undirected, utilitarian pseudo–design process which can easily be explained without invoking magic, or an intelligent, incompetent designer who contrives to appears to be as unintelligent as possible, who seems not to know what it did yesterday or why it did it and who has a plan that looks exactly like no plan at all.

Where is a purpose discernible in all this? There doesn't even seem to be an anthropocentric purpose to it all. There is massive complexity in the whole system and all for no ultimate purpose.

Heliconius **Butterflies and Mimicry.**

Heliconius butterflies are widespread in tropical and subtropical areas of the Americas, especially in the Amazon where they were the subject of a study by the biologist H.W. Bates [45] who gave his name to Batesian mimicry.

Batesian mimicry is where one harmless species mimics another poisonous or distasteful species, so avoiding being eaten by predators which mistake it for the harmful species. This not the same as cryptic camouflage where a species resembles its background, since this protection hides the species rather than deterring a predator.

For example one species of butterfly, *Heliconius erato,* can have a number of regional colour variations which are local to a given range. This species is distasteful to predators which recognise its markings as warnings. Another species, *H. Melpomene*, which is not distasteful, very closely mimics *H. erato* to the extent that it can be difficult to tell them apart and these also vary according to the geographical range of

H. erato, so there is no doubt that mimicry is happening and the similarity of markings is not just some extraordinary coincidence.

Examples of mimicry can be found throughout nature; moths and beetles can look similar to wasps and bees; harmless snakes can mimic harmful snakes.

A slight variation on this is known as Müllerian mimicry where both species are poisonous of distasteful so they tend to converge on as single set of markings. An example of this is the South American poisonous frog, *Pseudacraea eurytus*, which mimics different other poisonous frogs in their range, so *P. eurytus* varies across its range according to the pattern of its co–mimic species in different locations.

So, what was this putative intelligent designer thinking of here? Why design a prey species to feed a predator then redesign the prey so it deters the predator by tricking it? And why not just make the prey species poisonous or distasteful too? After all, it presumably did it once for the species being mimicked so why design such a complex different solution to the same problem? Solutions for problems that are themselves solution to other problems that were designed as solution… etc., etc., etc... and so *ad infinitum.*

None of this is a problem for evolution to explain; in fact it is predicted by it. In what way exactly is design by an intelligent, omnipotent designer a better explanation?

The final example of waste due to lack of planning and foresight from the butterfly world is that of the painted lady, *Vanessa cardui.*

Painted Lady.
The Painted lady butterfly is one of Britain's most popular butterflies, appearing in early April as is migrates into the British Isles from North Africa via Europe, reaching all parts of Britain, even the Orkneys and Shetland Isles. It is the only butterfly ever to be recorded in Iceland. Its

numbers vary hugely from year to year being abundant in some years and rare in others.

In Africa and Southern Europe it breeds more or less continuously throughout the year. In Britain, it begins breeding soon after arrival and, in a good year can produce two successful broods, so its numbers peak in mid–August as native–born butterflies supplement the migrants.

And, come winter, every single one dies.

No eggs, pupa or larvae survive even the mildest winter in Britain. The entire migration and breeding in the British Isles has been a colossal waste of time and resources. The entire British population will need to be replenished from Africa and Europe again next year.

This, if the ID lobby is to be believed is, all the work of an intelligent designer.

At the risk of labouring the point, I will turn now to another British migratory species, the European eel.

European Eel.

The European eel, *Anguilla anguilla,* hatches from eggs in the Sargasso Sea, off the coast of North America. It then drifts North Eastward on the Gulf stream as larvae known as a leptocephali until, 300 days later, they approach the European coast where they metamorphose into transparent 'glass' eels and enter the European rivers, changing from salt to fresh water and metamorphosing again into miniature adult eels or elvers.

They then spend anything from five to twenty years as yellow eels, feeding in rivers, lakes and canals and even sewerage systems, until they reach sexual maturity when they change again from the familiar yellow eels to silver eels with silver sides, a dark back and a white underside. During this maturation period they have been building a store of fat which will sustain them during the next phase of their life–

cycle, when they migrate down the rivers, across the North Atlantic and back to the Sargasso Sea. During this migration they don't feed and digest their own digestive system, living entirely on the fat store laid down in their earlier life. They will not be coming back.

In the Sargasso Sea they mate and lay eggs to produce the next generation and die.

The European eel is now classified as critically endangered due to a number of factors, some man–made but one being parasitism by a nematode worm, *Anguillicoloides crassus*, a parasite of the Japanese eel introduced to Europe in the 1980's. The nematode enters the eel's swim bladder and severely compromises its buoyancy and ability to swim. European eel numbers have declined by between 90 and 98 percent since the 1970s.

Apparently, the intelligent designer saw fit to design this migration from the Sargasso Sea and back so that a new population of eels can live in the rivers of Europe. Presumably, the readily–available solution of simply mating and laying eggs in the rivers in which they live, as most other fish and other organisms do, was not complex enough. What was required was a ludicrously long migration to a specific part of the Atlantic Ocean, for which the eel needed to take up to twenty years to prepare.

Of course, we have the characteristic prolific waste too, in that of the thousands of glass eels that make it to the European rivers, only one or two survive to migrate back to the sea to breed. There is no estimate of the fraction of eggs and larvae that make it to the glass eel stage.

And then the intelligent designer designed a nematode worm parasite to make the journey just about impossible for the adult eels.

There are very many examples of organisms both large and small which exist for no discernible benefit other than for themselves and even whole interdependent isolated ecosystems that could not by any stretch of the imagination be there for the benefit of mankind or over which humans could even be said to have any dominion – the alleged 'purpose' for which they exist according to creationists.

These are to be found in extreme conditions in environments hostile to human life such as around deep ocean 'black smokers' or hydrothermal vents, isolated and effectively sealed caves, and hot springs associated with seismic activity and volcanoes.

Deep beneath the Pacific Ocean, around vents in Earth's crust, are whole colonies of creatures that depend not on sunlight and oxygen but on geothermal heat and sulphur and the bacteria that can process these and tolerate extreme temperatures and pressures that would kill most other organisms in seconds. The chemosynthetic bacteria live close to the upwelling hot water bringing dissolved minerals including hydrogen sulphide from the magma below. Living off these bacteria are numerous species including Giant tubeworms – of which more in a moment – clams, limpets, shrimps and crabs.

Compared to the deep ocean floor away from these hydrothermal vents, where living things are entirely dependent on the 'rain' of dead organism which eventually make it to the sea bed, and where life is therefore sparse, the density of life around the vents can to 10,000 to 100,000 time higher, so rich is the nutrient supply provided by the chemosynthetic bacteria.

Giant Tubeworms.

The Giant tubeworms, *Riftia pachyptila,* living around hydrothermal vents have develop a unique method of obtaining their nutrients, and in doing so can sustain huge populations of individuals up to 2.4 meters (7 feet) long. In an example of evolutionary loss of complexity, these worms have no digestive tract and don't need to eat at all because they have evolved a symbiotic alliance with chemosynthetic bacteria that live inside the cells of their bodies and provide the tube worms with all the nutrients they want in return for protection and a supply of hydrogen sulphide, carbon- dioxide and oxygen from the surrounding water which is obtained through a red 'plume' which also serves to excrete chemical waste.

To reproduce, females release eggs into the water which begin to float upwards. The males then release sperm which swim towards the eggs. Fertilised eggs develop into free–swimming larvae which swim down and attaches to rocks. They are the fastest–growing invertebrates known, reaching 1.5 metres in less than two years.

I'm willing to bet that any ID proponent would be hard–pressed to come up with a purpose for these Giant tubeworms or indeed any of the other fauna associated with them, and certainly not in terms of their relevance to humans or indeed their dependence on humans for management services, or whatever 'dominion' over them means. The fall back would inevitably be some reference to mysterious ways or some other theological, non–scientific device, about the intentions of the intelligent designer not being knowable.

The unique conditions around these hydrothermal vents have led to them being the current favourite for the site where the first self–replicating systems, still too primitive to be called organisms, could have arisen in the earliest stages of the development of living systems in the process known as abiogenesis. I wrote about this in my book, *What Makes You So Special: From the Big Bang to You.* [46]. Since this is outside the scope of this book, I won't dwell on this further but interested readers can read the New Scientist article by Nick Lane and Michael Le Page [47]. Suffice it to say that abiogenesis is not the problem for science that the ID lobby would have you believe. There is nothing in the laws of chemistry and physics that prevent abiogenesis, given the right conditions and enough time.

Movile Cave, Romania.

A superb example of evolution in isolation can be seen in the Movile Cave in Romania at Constanţa near the Bulgarian border, which has been completely sealed from the outside world for about 5.5 million years. It is even sealed by an impermeable layer of clay from recent surface water as was shown by the absence of radioactive isotopes coming from the nearby Chernobyl nuclear power plant disaster which contaminated the area. The cave receives its water supply from 25,000 year–old water in neighbouring spongy sandstone formations.

The only food and energy source is the limestone rock the cave is composed of which is gradually being dissolved, so enlarging the cave, by the acidic carbon–dioxide atmosphere in the cave. This is used by chemosynthetic bacteria which form the base of the food chain with apex predators such as centipedes, spiders, water scorpions and carnivorous leeches living on the arthropods and worms that live on the mats of bacterial foam that form on the surface of the water. The species live in total darkness.

There are some 48 different species in the cave of which 33 are believed to be unique. One interesting example of how evolution can get stuck is the fact that some spiders still spin webs despite there being no flying insects. These catch the springtails which leap to escape predators and sometime land in a spider web. Many of the species have evolved very long antennae to compensate for the loss of sight in the complete darkness of the cave.

The inhabitants of this cave form a complete, self–sustaining ecosystem in isolation from the rest of the world and which could not possibly have been designed with humanity in mind. It appears to have no purpose outside the cave. Had it not been discovered in 1986 it might well have remained undiscovered until the biota had consumed the fabric of the cave itself – at which point it would cease to exist and no–one would be any the wiser. There may be other similar isolated caves and subterranean ecosystems of which we are entirely unaware; we just don't know about them.

Those then are examples of needless complexity, whether it is a species which appears to do nothing and has no purpose, not even an anthropocentric one, apart from making copies of itself, and in the case of mimicry, a more complex way to achieve the same result rather than using a method which has already been designed – and so available to be mimicked.

I will now look at the huge complexity found in most cells.

A single cell is a hugely complex structure in its own right. It is debateable, because no standard definition or measure of complexity exists, where most of the complexity of a multicellular organism is. Is it in the cells of which the organs are composed or in the arrangement of those organs? Are two complex cells joined together twice as complex as a single cell or does there need to be a degree of specialisation for complexity to increase?

Certainly, we know it took a very long time for simple cells like bacteria to evolve into complex cells like amoebae, paramecia and single–celled algae and simple fungi. Using an analogy Richard Dawkins uses in *Unweaving the Rainbow* [48], if you stretch your arms wide to represent the history of life on Earth from its beginnings at the tip of your left finger and ending today at the tip of your right finger, from its start across your midline to well below your right shoulder, it was nothing but bacteria. Multicellular life only flourishes by about your right elbow.

The consensus now is that complex cells or eukaryotes developed from symbiotic alliances between different bacteria or prokaryotes and that some of the organelles now found in cells may once have been free–living bacteria – the endosymbiosis theory, first proposed by biologist Lynn Margulis writing under her married name, Lynn Sagan [49] and initially rejected by mainstream biology. The mitochondria, common to almost all eukaryote cells, for example, still have their own DNA (mtDNA). The chloroplasts, which we will meet again when I discuss examples of poor design, were almost certainly once free–living cyanobacteria.

So, for something like two thirds of the time living things have existed on Earth, they weren't even prokaryote single cells but were bacteria. They had a long time and billions of generations to be filtered by the sieve of selection and honed to perfection by it. It was only when some of them got together in a symbiotic alliance – which might well have begun as a parasite–host relationship that eukaryotic life emerged. Then, relatively quickly – a mere few hundred million years – multicellular forms emerged to become the forms of life more familiar to us than the invisible to the naked eye bacteria and protozoa that still teem in the soils and water; in fact almost anywhere, both inside us and outside us.

It has been said, and biologically it is true, that we still live in bacteria world because eukaryotes and multicellular organisms are really just bacterial colonies doing what they need to do to survive and get through the selection sieve at each generation.

I will look now at some of the processes that almost all cells perform.

Mitochondria.

Mitochondria are the power–houses of the cell. Their function is to make adenosine tri–phosphate (ATP) out of adenosine di–phosphate (ADP) and phosphate. To do this they break glucose down into carbon dioxide and water in a metabolic pathway known as the citric acid cycle. This metabolic pathway is also knows as the tri–carboxylic acid cycle or the Krebs cycle.

ATP is used in multiple different metabolic processes by supplying the energy stored in the phosphate bond, so reducing it back to ADP and phosphate.

The citric acid cycle is so fundamental to living cells that it probably evolved very early in the history of living things. Some authorities have proposed that it may have arisen abiogenically before even recognisable simple cells had evolved [50] and was thus one of the processes that became organised within a membrane to form the earliest self–replicating systems, the precursors of proper free–living cells.

Mitochondria have their own DNA in a circular chromosome showing similarities with Rickettsia bacteria from which they are believed to have evolved. However, they have a greatly reduced genome – an example of evolutionary **loss** of complexity. They also have their own ribosomes – the structures where amino acids are built into proteins according to a DNA template.

From an ID perspective, was this complexity really the minimum necessary to supply the cell with an energy source? Well, perhaps surprisingly, but embarrassingly for the ID lobby, it would seem not. At least one group of single–celled eukaryotes, the *Monocercomonoides,* manages perfectly well without mitochondria and without any mitochondrial DNA, so their function hasn't simply been transferred to somewhere else in the cell. Admittedly, this group of organisms have a specialised environment, living in the gut of small mammals, snakes and insects but this shows that any supposed designer has designed an alternative that doesn't include a highly modified bacterium living in the cell.

But the conversion of ADP to ATP is a basically simple chemical reaction. Was it really beyond the wit of an intelligent designer to come up with a simpler method for making chemicals react? Ironically, the ID lobby point to processes such as the citric acid cycle as examples of irreducible complexity and hence, fallaciously as we saw earlier, systems that must have been created together. What they never question is whether they are needed at all. No intelligent, competent designer designs complex solution when simple ones could do the job.

Chloroplasts and RuBisCo.

The chloroplasts in the cells of green plants are, like mitochondria, examples of prokaryote cells which have become incorporated into cells to form eukaryote cells. In this case, the endosymbiotic organisms were almost certainly the descendants of the cyanobacteria that gave us one of the great mass extinctions – that caused by pollution of the atmosphere by their waste product – oxygen.

What these prokaryote cells had succeeded in doing was using photons from the sun as the energy source and the naturally–occurring inorganic molecules, water and carbon–dioxide, to build a short chain of organic carbohydrate – glucose – in the process known as photosynthesis.

While this gave these organisms a tremendous advantage the by–product of this reaction is molecular oxygen – a highly reactive oxidizing agent and toxic to most of those then living organisms that had evolved in an oxygen–free environment.

So living organisms were faced with an adapt–or–die situation; an intense selection pressure under which the survivors would inhabit an oxygen–rich world with new niches and new opportunities and the others would become evolution's casualties. Incidentally, this gives the lie to the frequent creationist claim that abiogenesis could not have happened because the oxygen in the atmosphere is corrosive. There was no oxygen in the atmosphere then. Oxygen is a by–product of photosynthesising organisms.

It is believed that early proto–plants may have been parasitised by these early cyanobacteria or maybe they were predated on by being engulfed by phagocytotic proto–plants, but that the relationship would have become quickly symbiotic, when the cyanobacteria were able to export spare glucose the early plants could use as their energy source. These symbiotic alliances became the earliest true plants – the algae.

Because of these chloroplasts in green plants we now live, outside the deserts and polar regions, on a green planet and most living organism are now dependent, directly or indirectly, on the glucose photosynthesising plants provide. The green of chloroplasts, or rather the green chlorophyll pigment they contain is abundant everywhere. Indeed, one of the component parts and an essential enzyme in the process for making glucose is probably the most abundant protein on Earth.

It needs to be. It is probably one of the least efficient and one of the best examples of an evolutionary blunder that can't be undone because there is no mechanism for scrapping the 'design' and starting again.

The enzyme goes by the scientific name ribulose-1,5-bisphosphate carboxylase/oxygenase, mercifully abbreviated to RuBisCo. RuBisCo is the enzyme in photosynthesis responsible for taking carbon dioxide (CO_2) from the atmosphere and building the chains of carbon in sugars which form the backbone of all organic substances.

But RuBisCo is incredibly bad at doing what it does; only carrying out about three reactions a second against tens of thousands of reactions a second for some enzymes. And it makes lots of mistakes. It finds it difficult to tell oxygen molecules (O_2) from CO_2 and often incorporates it by mistake, causing a chain reaction which causes a loss of carbon and wastes energy. To make matters worse, RuBisCo can end up making xylulose-1,5-bisphosphate which actually inhibits it! [51] Some plants have evolved mechanisms for reducing these mistakes – normally by keeping O_2 out of the way – but they appear to have evolved several times, independently and none of them are especially successful.

The origin of this problem is that RuBisCo itself evolved in an oxygen–free atmosphere, so the potential to make this mistake was not a factor natural selection could take into account. Unlike the way an intelligent, omniscient designer would work, natural selection acts on what is, and the here and now, not on what might happen later. Evolution can't pop into the future and see how things turn out and it has no reverse gear. But an omniscient, omnipotent designer would know in advance how things were going to turn out. And if it **did** make a mistake an intelligent designer would be capable of scrapping the design and starting again.

The Unintelligent Designer

No omniscient, intelligent designer would blunder blindly into a problem of its own making and then find itself stuck with the problem, unable to go back and start gain and having to make do with massive inefficiency and massive waste.

Having started off down the road to photosynthesis, and having given evolving life forms such a tremendous advantage, despite the inefficiency, there was no going back. Any tendency to change it would result in something even worse, so living things have to make do with what they have got. No planning; no ability to go in reverse, and no one to stand back and think of a better way, and start again. The fact that several plants have evolved different ways to compensate for RuBisCo's inefficiency shows that it not ideal for purpose. No omnipotent intelligent designer would come up with something which has to be compensated for.

On its own, RuBisCo, more than any other phenomenon in the natural world, dispels any notion of intelligent design.

What we now have is a world where masses of energy and resource have to be devoted to overcome the inefficiency of RuBisCo at producing something that almost every living thing depends on, because evolution headed down a route from which it could not turn back.

Just as we would expect, we also have massive needless complexity just to achieve the chemical reactions needed to convert inorganic molecules into a simple six–carbon sugar:

$$\text{Sunlight} \\ \downarrow \\ 6CO_2 + 6H_2O \rightarrow C_6H_{12}O_6 + 6O_2$$

This leads on to another example of needless complexity – massive tree trunks. Not examples of complexity within a cell, of course, but undoubtedly a consequence of it.

Tree Trunks.

Despite their usefulness to mankind as sources of fuel and building materials, biologically, tree trunks are the wasteful result of arms races that any intelligent designer would have avoided.

Because trees are dependent on two things – photosynthesis and the need to get their seeds pollinated - they have to struggle to get their leaves and flowers up above the cover produced by other trees with which they are in direct competition. The only way they can do this is to try to outgrow their neighbours by having taller trunks.

The tree needs to devote huge amounts of energy into making literally tons of the structural carbohydrates, cellulose and lignin, and all using the stupidly inefficient enzyme RuBisCo, to make the building blocks of these structural carbohydrates, glucose, just to stand still in the arms race.

Yet an intelligent designer could have achieved the same result by having all trees with very short trunks if any at all with none of the attendant waste. The one ID proponents advocate however seems to have gone into an infinite loop, designing solutions to the problems it designed as solutions to earlier problems, *ad infinitum*. Masses of waste and all to no discernible ultimate purpose.

But here again, this putative intelligent designer also designed herbivorous animals to eat the product of the trees' efforts, so having them all at ground level would have created another problem for it to solve – how to prevent the herbivores doing what it designed them to do.

And what was its solution? Why, another wasteful arms race, of course!

So we have very tall acacia trees growing isolated on the African plains with no competing neighbours stealing their sunlight and hiding their flower from pollinating insects; and we have giraffes with ludicrously long necks with a quite ridiculous recurrent laryngeal nerve pathway down to the chest cavity and back up again just to get from the brain to the throat so the giraffe, like all mammals and birds can cough. They need to cough because of another design blunder – the airway crosses the food path so food can get into the airway, causing choking.

The giraffe also needs special valves in its neck blood vessels so it can drink without having a brain haemorrhage and stand up again without losing consciousness from postural hypotension, and all to overcome the designer's 'solution' to the problem of it doing what it was designed to do and eat acacia leaves – apparently.

The last example of needless complexity I intend to look at is the complexity of the genome itself, in particular the vast amount of redundant or 'junk' DNA we see in every genome of every species for which we have a complete genome analysis.

Junk DNA.

Redundant or 'junk' DNA is that DNA in a species genome that seems to do nothing. It is the DNA that doesn't code for proteins or do anything else. It doesn't even mark the end of a length of coding or the end of a chromosome.

It just sits there, getting replicated, sometimes mutating, generation after generation, in every cell of every organism. It gets copied in the gametes and joins other copies of other junk DNA in the zygote, using up the resources and energy and so imposing a cost on the organism, all for nothing.

By careful analysis, scientists are able to show that origins of this junk. Some of it is due to accidental duplication of lengths of DNA at some point in the organism's evolutionary history; some of it comes from ancient retroviral DNA. By looking for the same junk in other species, these lengths of junk DNA can be seen to have been present in common ancestors, so are strong evidence for common origins and for descent with modification. Some of the retroviral DNA, for example, probably became inserted in the remote fish ancestors of humans since they are to be found in all the descendants of the lobe–finned fish that crawled onto the land and became terrestrial tetrapods from which all amphibians, reptiles, birds and mammals evolved.

Retroviruses have a sneaky way of disguising themselves when they infect a cell. Although they are RNA viruses, they insert a DNA template for their RNA in the host's own DNA, where the host's immune system can't distinguish it from the host's own DNA. This DNA is replicated along with the normal genome into which it has now become incorporated. One example of a retrovirus currently infecting humans is the Human Immunodeficiency Virus, HIV which causes AIDS. Over time, retroviruses and their hosts coevolve so that the host is no longer harmed, at which point the retrovirus becomes junk DNA and can mutate and even be co–opted for novel uses.

It has been estimated that between five and eight percent of the human genome is composed of retroviral DNA. This is not especially unusual either; it is true for all jawed vertebrates [52].

Junk DNA is a major embarrassment for ID advocates because it is simply not intelligent to create and repeatedly replicate DNA for no discernible purpose, so they spend a great deal of effort trying to find evidence that all DNA has a function. For example, and to illustrate the lengths they will go to, see one such anonymous attempt by *Answers in Genesis* (AiG) [53]. AiG is creationist Ken Ham's online vehicle; the article attempted to present the finding by a Cornell University research group [54], that junk DNA in two closely related species of fruit fly (*Drosphila*) may contribute to inhibiting interbreeding, so maintaining speciation, as supporting Young Earth Creationism (sic).

Really? A supposedly intelligent designer came up with **that** solution to the problem of two of its creations interbreeding? And that proves Earth is just a few thousand years old! Presumably, this is a prime example of 'Creation Science'. If it is not a prime example of needless complexity it would be difficult to think of one. It is not as though this effect of junk DNA is normal. This was a discovery worth writing up and publishing in a peer–reviewed science journal.

Another measure of the concern junk DNA causes creationism was their response to a highly–criticised [55] ENCODE (Encyclopedia Of DNA Elements) study that, on a very liberal definition of 'functional' which some people have described as so wide as to render it meaningless, came up with a figure of 80 percent of the human genome being functional [56]. **Any** evidence of activity was regarded as functional even if no function as such could be identified.

Creationists greeted this with barely–concealed ecstasy [57] [58], ignoring the fact that even that study still left 20 percent of the human genome as junk, and of course ignoring subsequent studies that came up with a much lower figure for the percentage of functional DNA [59].

In the next Chapter I will use the example of the Norwegian spruce – the Christmas tree as an example of prolific waste. I could equally have used it here as an example of needless complexity. It has so much junk DNA that its genome is seven times bigger than a human genome yet it has less than double the number of functional genes. The problem is that an error–correcting system which is meant to correct accidental duplication is itself broken. Yet Norwegian spruce do not seem to suffer any serious consequences. The massive volume of junk DNA is just replicated *ad infinitum* in trillions of Christmas tree cells for no reason at all.

Also from the plant world, we have the example of the Night-flowering catchfly, *Silene noctiflora*, but this time it's not the nuclear DNA but the mitochondrial DNA that contains huge amounts of junk.

The mitochondria in the Night-flowering catchfly have more than 50 circular chromosomes totalling more than 7 Mb of information. This number of chromosomes can vary by 19 entire chromosomes between populations. These 19 chromosomes either have no genes at all or have duplicates of genes found in other chromosomes, so they appear to serve no purpose at all and some populations manage perfectly well without them. However many chromosomes an individual *S. noctiflora* has in its mitochondria, it only uses 54 genes which comprise only a tiny fraction of this massive and variable genome.

Perhaps not surprisingly, you will not find any references to the mitochondrial DNA of the Night-flowering catchfly in creationist literature.

Creationists will argue that, because some non–coding DNA seems to have some regulatory functions, it is not junk at all and is there by design. If this were true, we would expect there to be about the same amount of non–coding DNA across all genomes. The problem is, this is not what we see. The proportion of non–coding DNA varies hugely. In at least one species of plant, the Carnivorous bladderwort, *Utricularia gibba*, junk DNA comprises just three percent of the plant's unusually small genome [60]. This plant appears to thrive without the supposedly functional non–coding DNA. This suggests that whatever functions non–coding DNA might have in some species, it has either been co–opted for that purpose or the function is non–essential or can be achieved other ways.

The Unintelligent Designer

7. Prolific Waste

We have seen in previous chapters how wasteful natural systems can be, especially when it comes to reproduction where little is left to chance when it comes to avoiding failure. We saw, for example, how parasitic worms produce thousands of eggs or larvae in order to end up producing a few adults and how the malaria parasite needs to produce millions of gametocytes in order to ensure a few are taken up by a feeding female mosquito.

We saw how Painted lady butterflies migrate from Africa and Southern Europe each year into Britain and Northern Europe in their hundreds of thousands, even millions to breed throughout the summer, only for every single one to die because they cannot tolerate even the mildest of winters in northern climates. We saw also the lengths European eels need to go to breed and produce vast numbers of larval eels to ensure a few make it back into the rivers to live and breed the next time.

And we saw how probably the least efficient enzyme known needs to be present in astronomical numbers in a green plant in order to produce the sugar by photosynthesis that almost all life depends on, save for those living in the remotest parts of the planet.

I'll look now at a few more examples of prolific waste.

Giant Puffballs.

Giant puffballs, *Calvatia gigantea*, can be a foot or more in diameter and sometime over three feet. Some have even been mistaken at a distance for sheep. They exist to produce more puffballs. If chosen very young they can be cooked and eaten but are dangerous to eat when the spores form. Nothing else eats them.

They spend most of their life as fungal hyphae in the soil until, when the conditions are right, they will produce the large fruiting body familiar as the puffball.

Spores are produce inside this fruiting body, the entire inside being converted to the greenish yellow mass of spores. Each fruiting body can produce 70 trillion spores. Only one or two of these per year need to become a new puffball for the population to remain stable.

If every spore became a new puffball their spores would exceed the number of suns in the known Universe and their combined weight would exceed that of Earth.

And this is to produce one or two mature individuals.

ID proponents have concluded that this is the work of a highly intelligent designer.

Pollen Grains.

At the bottom of the close I live in is a grassy playing area with a few trees to one edge. One tree in particular is a small Oak. In late spring this Oak produces several tens of thousands of long racemes of small male florets, each raceme composed of maybe a hundred florets so the small tree will have maybe a million male florets.

Each of these florets will produce hundreds, maybe thousands of pollen grains. This single small tree probably produces hundreds of millions, maybe a billion or more pollen grains, each capable of fertilising a female flower to produce an acorn if it happens to get blown by the wind in the right direction. The tree will then produce maybe a few thousand acorns, each with the potential to become a new Oak tree.

To maintain a stable population a single Oak needs to produce just one successful new Oak tree during the whole of its lifetime of maybe several hundred years.

To achieve this, it produces a billion or so pollen grains and some tens of thousands of acorns every year of its life and devotes the whole of its resources to that effort. It grows a massive trunk and branches to get its leaves up into the sunlight so its trillions of chloroplasts with their hugely inefficient RuBisCo that we met in Chapter 5, can make the sugar it needs to build the wood it needs and grow the leaves and flowers it needs. It needs a water transportation system to pump water and minerals up to the canopy to supply the leaves. Its roots need to dig deep into the soil to get them, aided by symbiotic fungi. The energy needed to raise that weight of water to the height of even a small tree, like that at the bottom of our close can be considerable.

And all to produce a single successful acorn in its lifetime that grows into another Oak tree.

This is just one example of the waste that reproduction by wind–blown pollen produces in most grasses and many trees. I have seen a small ornamental cedar with red male catkins almost appear to be smoking as clouds of red pollen were rising from the top of it in a light breeze. In spring on a dry day, you can put a piece of adhesive plastic in the open, sticky–side up, for a few hours then count the pollen grains under a microscope. The air is full of these microscopic male plant gametes, each with half the DNA needed to make a whole plant when united with the other half in the female gamete.

Yet only a tiny fraction of a fraction of one percent of them will ever do so.

A similar situation with waste in reproduction exists with most animals where millions of sperms are produced just to fertilise one egg.

Eggs and Sperms.
This situation is particularly difficult for creationist to explain because central to their belief system is the belief that a deity creates at least every human specially, fully intending to produce exactly that human.

The Unintelligent Designer

You and I, and every one of us are supposedly created with us in mind. Indeed, how could it be otherwise with an omniscient deity?

And yet maybe half a billion sperms are directed at a single egg when we are conceived.

If the intention was for a preselected sperm to be first to reach the egg and fertilise it, what were the others for? If it could really think of no other way to create humans than with eggs and sperm, why did this putative designer not design it so just one sperm, the exact sperm needed with exactly the right DNA, was produced and have a system guaranteed to ensure it reached the egg? This system should have been well within the capabilities of an omnipotent deity.

Instead, it seems to rely on chance, not only over which sperm gets to the egg but whether there is an egg to be fertilised. The whole process is dependent on so many unpredictable variables such as exactly when the parents had sex and even in what position they did it, and maybe on what they did soon after. And after all that, there is the half a billion to one chance of that sperm that gave you half your DNA being the one to win the race and there is a 50:50 chance of you being male or female – that most fundamental thing about you.

Creationists believe this system was intelligently designed to ensure a predetermined outcome, every single time.

The reason so much waste is seen associated with reproduction in both animals and plants is because of the way evolution works. Evolution is nothing more than the process of producing the most copies of the different alleles of the same genes. The allele that produces the most copies of itself compared to other alleles is the winner. If it produces more copies of the allele that causes Oak trees to produce a trillion pollen grains a year than one that produces a mere billion copies then the trillion pollen grain producing allele will come to dominate in the species gene pool. There is no mechanism for standing back, calling a truce and having all Oak tree downscale their pollen production.

But of course an intelligent designer could do just that even if, despite its omniscience, it had failed to understand what a wasteful process this was becoming.

The next example could equally be an example of needless complexity but I've used it here as an example of prolific waste because it appears to use resources for no discernible benefit.

It is also an embarrassment for creationists who assume that the complexity of an organism can be measured by looking at the complexity of its DNA, stemming as that does from the assumption that somehow DNA is a blueprint or plan for making whatever species it makes. Of course, with humans sitting at the pinnacle of creation, according to the creationist anthropocentric view, and with complexity being seen as a measure of the creator's skill, we should expect the human genome to be the most complex in the animal kingdom.

It comes as a shock to many creationists that this is far from the truth. In fact, the human genome is about average for a mammal and much smaller than some creatures such as amphibians like the axolotl and even the humble Christmas tree or Norwegian spruce.

The Christmas Tree.

The Christmas tree or Norwegian Spruce, *Picea abies* has a little surprise (or maybe that should be big surprise) for creationists and it's not wrapped in festive wrapping.

The tree has a massive genome, far larger than that of a human being, yet no–one could seriously suggest the Christmas tree is far more complex that a human being. This was discovered by a research team from the Umeå Plant Science Centre (UPSC) in Umeå and the Science for Life Laboratory (SciLifeLab) in Stockholm, Sweden, led by Björn Nystedt and published in *Nature* in 2013 [61]. Parts of this item are taken from my blog post written to draw attention to this research [62].

The genome is seven times larger than the human genome and contains about 29,000 functional genes compared to the 17,000 functional genes of the human genome. So, it seems the Christmas tree requires a genome seven times as large as humans to contain less than double the number of genes.

How did this huge genome come about?

One way, common in plants, by which a genome can be increased in size is by whole genome doubling where a mistake in the production of the reproductive cells produces pollen or ovules with the full complement of chromosomes instead of the normal half set. If this is fertilised with a similar diploid gamete it can result in a tetraploid version. The Cox's Orange apple is a tetraploid apple, for example. Another way is a simple doubling of a length of DNA during its replication, so the same length is replicated twice and becomes incorporated in the normal genome.

A third way is by using a crude, near enough is good enough approach to replication. Imagine you are a computer programmer who needs to write a routine for copying a table of data but the size of the table can vary so you don't know how big the table is. To save time and because it is easy, you write a routine to copy a chunk at a time until you are well past the end of the table and just leave it at that. Much easier than including a routine to work out the exact end of the table and so the exact length of the last chunk to be copied.

You copy the whole table and a lot of following junk as well. Next time the routine runs it copies everything you copied earlier and a whole lot of new junk as well. So far as the user is concerned, the table is there and all seems okay, until eventually the memory footprint of the application gets massive because your sloppy, near enough is good enough routine has filled up the hard drive. Not good programming, but evolution isn't bothered about the future because it can't plan, and the analogy between a computer storage devise and DNA falls down at that point because DNA is its own storage medium. Near enough is good enough works for evolution because evolution is unplanned and utilitarian. As a piece if intelligent design however, it's stupid. The programmer should have been thrown off the programming course.

The pines are gymnosperms but the scientists who carried out this analysis point out that there is no such evidence of gene doubling in the gymnosperm lineage. The only feasible explanation is a gradual accumulation of mostly redundant DNA due to a faulty replication and a defective correction mechanism which, in other plants, helps correct this faulty mechanism. It's a normal feature of DNA replication in both animals and plants that the ends of DNA sequences are often replicated several times. These are replicated again in the next generation and, over time would lead to a huge amount of redundant coding. In most species this tendency is corrected and it is this which seems to have failed in the Norwegian spruce leading to this accumulation of DNA over about 200 million years.

Now, the obvious problem for creationists is how to fit this into their ID model – apart from it needing 200 million years to accumulate. Creationism relies on perpetuating several myths about the science of evolution because these myths or parodies are what creationist pseudo-scientists earn their living attacking.

One of the myths is that biologists think that evolution is about how human beings came about and that the entire point of evolution has been to evolve human beings. This means that human beings have to be presented as the most highly evolved and most complex of species, at least in the creationist view of evolution. This is, of course, nonsense since evolution is all about how diversity arose and it has no aim or objective. No single species can be said to be more highly evolved than another since all living species have been diversifying for the same length of time. Never-the-less, if you go to a creationist website you'll see that parody of evolution attacked time and again.

Another myth is that evolution always involves increased complexity and new information. This makes it easier to attack evolution with a scientific-looking claim that it somehow contradicts basic laws of physics such as the Laws of Thermodynamics and some half-understood dogma that no new information can ever arise because that contradicts some fundamental law related somehow to thermodynamics too.

This too is nonsensical and based on a deliberate misrepresentation of the Laws of Thermodynamics, which neither preclude a local decrease in entropy nor prevent new meaning to existing information arising if the environmental context changes, or new information arising for that matter. If it did there could be no life because chemical processes could not occur, nor could automobiles work or any wealth ever be created by doing work.

Thirdly, creation pseudo-scientists present an increased complexity in structure as reflecting an increased complexity in the genome. This would mean the more complex an organism, the more complex its DNA must be. This plays to the myth that somehow DNA is like a computer program, so more complex output must come from more complex input.

This is also nonsense because DNA is more like a recipe than a construction manual. Complexity can come simply from switching controlling genes on or off at different times, hence the human genome has almost the same number of functional genes as chimpanzees and gorillas and only differs from that of a mouse by about ten functional genes. The same ingredients; slightly different recipe.

So, creationists are faced with several dilemmas here:

- How do they explain a manifestly less complex organism like a Norwegian spruce having such a vastly more complex genome and almost twice as many functional genes as humans? If their parody of evolution is correct, humans would have the most complex genomes.

- If additional DNA means additional information, what new information is there in the redundant DNA in *Picea abies* and why does a spruce need seven times the information that humans need?

- How do they explain a species which diversified from the last common ancestor shared with humans about 500 million years ago having a more complex genome than humans? If their parody of evolution is correct the human genome should be the largest because humans are allegedly the most highly evolved of all creatures.

- How do they explain such a huge genome with so much redundant DNA? Why would any intelligent designer create so much redundant DNA?

- How do they explain a faulty DNA replication mechanism which needs an error-correction method to prevent it running out of control in the first place, and why would an intelligent designer then break the correcting mechanism it designed to compensate for its earlier mistake?

In addition to needless complexity, the humble Christmas tree is a manifest example of incompetent design and waste in nature. A great deal of effort and resource goes in to replicating masses of redundant DNA not just when the species reproduces but in every single cell in every single Christmas tree.

I will turn now to another example of prolific waste, Needless complexity and sheer bad design as revealed by the relatively new and still developing biological science of epigenetics.

Epigenetics.

Epigenetics is a relatively new branch of biology which is presenting biologists with some new challenges, not the least of which is its complexity. It is also presenting creationism with some exciting new gaps in which to try to insert a creator and attack science on the basis that if science doesn't know something it has failed and proven itself unreliable. No matter that in a few years, maybe a little longer, science, using the scientific method will one day close those gaps and, like all the other gaps, will fail to find a god in them. By then, creationist apologists will have moved on to some more gaps, still confident that **this** time science will be forced to admit that 'God did it!'

Whole books are being written about epigenetics and the challenges it presents to biology, one such being Nessa Carey's *The Epigenetics Revolution: How Modern Biology is Rewriting Our Understanding of Genetics, Disease and Inheritance* [63].

Briefly, epigenetics is the study of how various genes are turned off in different cells as they become specialised during embryonic development. The actual mechanism, the epigenetic processes itself, are beyond the scope of this book but what interests us here is what this means for our understanding of evolution and, more importantly, why the processes are needed.

The interesting thing from an evolutionary perspective is the fact that, when DNA in the form of a chromosome is replicated in cell division, the epigenetic 'settings' are also replicated, so any genes which were switched off in the parent cell will also be switched off in the daughter cells. It seems that, as cells differentiate and become specialised to perform specific functions as the embryo develops, genes can be switched off but they don't get switched on again, so cells derived from one layer will have all the epigenetic settings of that layer, and some more. The descendants of those cells in turn will have all those settings plus some of their own. And so on until in the fully formed individual all the specialised cell types will have their unique set of epigenetic settings, so liver cells will have different active genes to, say, muscle cells or brain cells.

This is of interest to evolutionary biologists, and probably why creationists get so excited by it, because it seems to imply that something acquired after the cell formed can be passed on to descendants, contrary to what Darwin argued and what current evolutionary theory says; that characteristics acquired after birth (or more correctly, after conception) can't be inherited because we get all our inherited traits from our parents.

This Darwinian view was in contrast to the rival theory proposed by the French biologist, Jean-Baptiste Lamarck, who argued that characteristics acquired after birth could be inherited. For example, a blacksmith's skills and his strong arms could be inherited by his sons. He suggested that giraffes could have acquired their long necks by stretching them to reach branches, so their offspring would inherit these longer necks.

Both Lamarck and Darwin knew nothing about how we inherit from our parents of course because they knew nothing of genes, chromosomes or DNA.

So, to a creationist, it might look as though epigenetics falsifies their arch–demon, Charles Darwin, so, in their view which sees 'Darwinism' (that is, the whole of evolutionary theory) as a dogma, this falsifies the entire body of science. Curiously, they forget that Lamarck's idea was simply an attempt to explain the observable fact of evolution. The reason Darwin's ideas became accepted is because they better explain the observable evidence in nature that living organisms can be arranged into branching hierarchies which show a clear relationship between them.

But, there is a world of difference between the individual cells of our bodies inheriting epigenetic settings from their parent cells, and whole human beings inheriting all the epigenetic settings from their parents. The obvious difference being that we as individuals don't inherit all our parent's cells; we inherit half our DNA from each parent via two gametes which fuse to form a single cell or zygote from which all our cells are derived.

These gametes are of course themselves the descendants of specialised germ cells and are specialised cells in their own right, but there is no reason to suppose that they will carry epigenetic settings in, say, brain or muscle cells. The resulting zygote needs to be pluripotent, that is having the capability to give rise to all the different cell types in the resulting adult, so it must have its epigenetic settings reset before any cell differentiation can take place. This resetting serves to remove any epigenetic changes acquired after conception by either parent, or at least the vast majority of them.

Having said that, there do appear to be a few instances where things that have profound effects on parents during their lifetime can influence their children and even grandchildren. One such, cited by Nessa Carey [63], is the results of the 'Dutch Hunger Winter' when, due to German blockades between November 1944 and spring 1945, food supplies were almost non–existent to the extent that some 20,000 people died of starvation. Studies on the children of this population, whose mothers were pregnant during this period, show that their subsequent development was affected by when during their pregnancy their mothers were under–nourished. This in itself is not surprising, because, although it was acquired after conception, it is not difficult to imagine how it could have affected epigenetic settings.

What was surprising, however, was that there seems to have been an effect on the children of these children. In other words, the grandchildren of the women who were starving during pregnancy could somehow have inherited something depending on when during their pregnancy their grandmothers were suffered a period of starvation [64].

Now this is something science needs to explain. How it fits in with the ID model is also something creationists need to explain. Simply waving it as an example of how Darwinian evolution might not be the whole story, as though that destroys the whole of evolutionary biology and negates the masses of evidence on which neo–Darwinian consensus is based, is grasping at straws.

But there is something else that the ID movement needs to explain and this brings me back to the point of this book in general and to this chapter in particular. Why is the whole complex process of epigenetics necessary in the first place?

Epigenetics is necessary because in multicellular organisms, any advantage of multicellularity is only realised by specialisation of cells and their arrangement into organs carrying out specialist functions. Many of these functions are only necessary in the first place because of multicellularity, of course. Mammals need digestive, respiratory and circulatory systems to get oxygen and nutrients and remove waste to the cells too far removed from the surface to do it the way single–celled organisms do it – by direct exchange with their environment.

Evolutionarily speaking, multicellularity gave some organisms an advantage over others, but it comes at a price. One of the prices is the complex system of epigenetics and the need to reset and start again with each new individual.

But why is this needed anyway? It is needed because, just like single–celled organisms, the cells of multicellular organisms inherit **all** the DNA of their parent cells regardless of their eventual function as specialised cells in specialised organs. Unless the unnecessary and unwanted genes are turned off there would be no specialisation and so no benefit from multicellularity. The last thing you want is your brain cells producing the digestive enzymes your pancreas secretes or your kidney cells producing the contractile proteins in your muscles. You want your cells to be specialised and be good at doing what they are specialised to do – and nothing else. When cells start becoming generalised and doing other things they are called cancer.

So what any ID model needs to explain is why any intelligent designer would arrange it so that all cells (with one or two limited exceptions) contain all the DNA of the entire organism when they only need a few special genes to function? Why is this complex system of epigenetics necessary in the first place? Why would an intelligent designer not design things so that as cells become specialised, they only get the DNA they need?

Instead, we have the ludicrous situation of prolific waste of resource in replicating all the DNA – with its attendant risk of going wrong – to have most of it permanently switched off in almost every one of our 70 trillion cells. Then we need a mechanism for resetting it and starting again in the newly–fertilised zygote.

In epigenetics we have a few exciting challenges for biology; for creationism we have as good an example as you can wish for of designer incompetence. We have prolific waste, needless complexity, a clear failure to plan ahead and needing to make the most of a bad job, and of a ludicrously complex 'solution' to a problem of its own making because, apparently, the designer lacked the wit to rethink the problem and start again.

How this can be described as intelligent design is quite beyond me. It requires definitions of 'intelligent' and 'design' that are unrecognisable and indistinguishable from the normal definitions of 'gross incompetence' and 'stupidity'.

For evolutionary biology, of course, epigenetics is as nice an example as you could wish for of the utilitarian, pragmatic nature of evolution, where natural selection can only act on the here and now and where any solution, no matter how suboptimal, will be adopted it if gives an advantage. It is an example of how, like the example of RuBisCo in Chapter 5, evolution has no reverse gear and cannot scrap a suboptimal solution and start again with a better one, as any intelligent design process should be capable of.

Living multicellular organisms are now stuck with the complexity and waste of epigenetics because that gave an advantage, despite the inefficiency and waste, of multicellularity over single–cellularity for some, but by no means all, organisms. Very many organisms remain single–celled of course, and very many remain prokaryote rather than eukaryote. Evolution does not have a plan and is not trying to achieve anything.

It might well be that the reason life on earth, using Richard Dawkins ' outstretched arms analogy [48], took until well past our right shoulder and down towards our elbow to even evolve beyond bacteria was that multicellularity and cell specialisation was not something that arose easily.

This would not have been a problem for an omnipotent designer, of course.

The Unintelligent Designer

8. Atavistic Genes.

Atavistic genes are those genes which are still present in the organism's genome but which are not expressed or which express in the developing embryo only to have the structure destroyed. These genes always code for something that was present in an evolutionary ancestor but which has since been lost. The genes do not switch on or they are deactivated or other genes are activated to undo their work.

Atavism is the reappearance of ancestral features due to a fault where the replacement gene fails to suppress the atavistic one or the atavistic gene overrides the new one [65].

Several examples have been revealed in birds, in addition to the very occasional hens' teeth. One of these was discovered by a Florida University–based team who solved one of the mysteries of bird evolution – how the males lost their penis [66].

The Cockerel's Lost Penis.

Most birds mate with a 'cloacal kiss' during which the male deposits sperm into the female cloaca without penetration. A few birds however, such as the ducks, geese and swans (Anseriformes) **do** have a penis, so clearly either a penis was present in a common ancestor or a few birds have evolved one.

By comparing the development of members of the Galliformes, which includes domestic hens, with that of Anseriformes the Florida University team found that a penis starts to develop in the embryo of the Galliformes in the form of a genital tubercle which in Anseriformes develops into a penis. But then a gene, *Bmp4* is activated in the tubercle, leading the cell death and failure to develop. They found that when they blocked the activity of this gene, a penis developed and when they activated it in the Anseriformes they failed to develop a penis.

In other words, cockerels have the genetic equipment to grow a penis but the cells which should express those genes are killed soon after they start to grow into one. This equipment has been replicating in every chicken cell and the cells of every related species, and probably almost all other birds since they diverged from the common ancestor shared with the ducks, geese and swans. All for nothing. Rather than simply deleting these genes from the bird genome, as any intelligent designer would do, evolution has countered them with a crude but effective gene which kills off their effort.

Dinosaur–Faced Chicks.
In May 2015 a Harvard–based team found that by blocking just two genes in developing chicken embryos, the chicks failed to develop a beak but developed instead what looked like dinosaur snouts [67].

Chickens still have the genes for growing a dinosaur face from their remote velociraptor ancestors, but these are overridden by two genes that instead grow a beak.

Would it be labouring the point to ask how keeping genes for growing a dinosaur face and then overriding them with genes for growing a beak instead is good, intelligent design? There are so many examples of these genetic fossils they would probably fill several large volumes if collected into a single set of books.

Snake Legs
In October 2013 a large team of scientists based at the Lawrence Berkeley National Laboratory, Berkeley, California found that snakes have the potential to grow limbs but that over the course of evolution, they progressively lost the function of an 'enhancer' gene that is essential for normal limb development [68]. Without this enhancer gene other vertebrates also fail to grow full limbs. Utilitarian evolution, oblivious of future cost and waste, simply achieved limbless snakes by

breaking one of the genes involved, but would an intelligent designer be oblivious of future costs?

In the same month, two researches at the University of Florida found that not only do python embryos develop the beginnings of hind limbs but that fossil evidence that some ancient snakes re–evolved limbs was due to these atavistic genes being reactivated. [69].

Well, maybe you **could** at a stretch make out a case for an intelligent designer keeping the means to make legs, planning to use them again later, but does that really make sense to copy those genes in every cell in every snake so that one day they could be reactivated? Why not simply wait and then insert the genes anew?

Because evolution doesn't plan. Evolution just makes use of what it has. In the environment in which growing limbs gave an advantage, growing an atavistic leg was an advance over limblessness and so was naturally selected for. In a different environment, it was an undesirable mutation that natural selection removed.

Human Tail.

To the eternal embarrassment of creationists who deny a common ancestry between humans and monkeys, humans are very occasionally born with a tail [70].

Humans **do** have a vestigial tail anyway, known as the coccyx – a short length at the end of their vertebral column consisting of 3–5 small bones, the vestigial remains of vertebrae, at the end of the sacrum. This doesn't show externally but can be felt. Occasionally, however, children are born with an external tail covered in skin and containing muscles, nerves, blood vessels and other soft tissues but not bone or cartilage. The cause of this is a failure in the system that stops a tail developing fully in the embryo.

According to the ID model, an intelligent designer designed humans with the ability to grow tails, using the same method that monkeys use; then it designed a method for stopping them growing a tail, using the same method as that for all the great apes, because **they** can grow tails too, but that method doesn't always work.

It is quite astounding that in the 21st Century, biologists are still attacked and pilloried for not believing in magic creation, while 38 percent of American adults think it is the best available explanation and that scientists are either mad or evil for not agreeing.

Dolphin Legs.

In November 2006 Seiji Osumi, an adviser at Tokyo's Institute of Cetacean Research, told a press conference that Japanese fishermen had caught a dolphin with well–developed, symmetrical rear flippers in addition to the normal front pair.

Whales and dolphins share a common four–legged terrestrial ancestor with hippopotami and deer from which they diverged about 50 million years ago.

These and the many other examples of atavism, such as additional nipples in humans, the dew claw often seen in dogs, additional toes in horses and guinea pigs and adult axolotls with webbed feet are biological curiosities that give a clue to the ancestry and ancestral phenotypes of the species. Similarly, the presence of 27 of the 29 genes needed for meiotic division in the asexually reproducing *Trichomonas vaginalis* that we met in Chapter 5 is evidence that it once had an ancestor that reproduced sexually. It can't possibly be evidence of intelligent design.

To a creationist and especially an ID advocate, they are anathema. They go to extraordinary lengths to mock the science, cast doubt on the evidence and slight the scientists who report them. With the irrefutable photographic evidence, they are difficult to deny outright but what they will never do is explain them biologically. Why would an intelligent designer design any creature with the ability to grow atavistic structures?

That's probably enough about atavism. There are only so many ways to make the same point.

In the final chapter, I will look at the question of purpose in design.

The Unintelligent Designer

9. Ultimate Purpose

Recalling back to what I argued in Chapter 1 that good design is always for a clear purpose, I will look now at how well living systems serve that purpose. Recalling also how in Chapter 6 I tried to discern what purpose Christians, Muslims and members of other faiths assume to be the purpose of their lives. This was needed in Chapter 6 which was about needless complexity as something to try to assess necessity against. Without knowing what the ID movement's putative designer is trying to achieve this would be difficult.

The problem is, what I discovered was rather nebulous; the purpose of human life is supposedly to worship the creator and obey all its rules so that there will be some sort of reward at the end of it. At different times throughout history, the actual purpose of particular human lives might have been to perform some allotted task such as being a humble labourer, a serf, servant or slave. In the case of women, a housewife, compliant and obedient sex toy and handmaiden to her husband, brothers and father; and of course, to be a mother, often with a Biblical or Quranic justification, the better to serve the needs of the ruling classes whose lives were deemed to be the only really worthwhile ones.

According to the Abrahamic religions, all the rest of creation is there to serve the needs of humans while they fulfil their allotted purpose.

And the whole thing; the entirety of creation, so that the creator will be loved, worshipped and obeyed. All of human activity and endeavour; the whole of civilisation; of agriculture, production of goods, of trade and economic systems; of organised religion and government was so humans could live to worship and obey gods and obey their rules, with the ultimate reward being left to their final judgement.

But why does that require a complex organism at all?

The Unintelligent Designer

Why could a shapeless, formless blob of amorphous matter not love and obey its designer if its omnipotent designer so designed it? Would that not have been the simplest of possible, minimally complex designs, perfectly fitted for its one purpose?

Of course it would; so why the hugely needless complexity?

Why the absurdly prolific waste? Why the arms races and parasites and organisms that apparently do nothing other than produce copies of themselves? Why does this intelligent designer spend so much time trying to solve problems of its own making, only to then treat its solution as a problem to be solved? How exactly do the arms races between bats and moths or the lives of Slug mites and Giant tubeworms deep in the Pacific mid–ocean trenches, help humans worship their putative creator?

It might be possible to argue with some superficial success that the wonder and breath–taking beauty in the world could be attributed somehow to a loving creator wanting to make a beautiful, even perfect, planet for its favourite creation, mankind, to live on, but how that debases it all! To wave it all aside as a conjuring trick makes us satisfied with not knowing how things really work and why they are really the way they are.

And it is simply not true. This is not a perfect place despite its beauty; far from it. Large parts of it are hostile, even lethal to human life and to any life not specifically adapted to the hostile environments. It is subject to volcanoes and earthquakes; to tsunamis and droughts; to hurricanes and tornadoes and occasionally to cosmic accidents such as meteorite strikes. With our modern clothing and buildings and the modern conveniences it is easy to forget that few of us outside the tropics would survive the winter naked and without shelter over much of our range.

This planet is not even the best of all possible planets. If a creator god had designed Earth, it could have done a far better job of it. It could have given us a sun which will not ultimately destroy us all when it becomes a red giant in about 5 billion years. In 2014, René Heller and John Armstrong of McMaster University, Hamilton, Ontario, Canada showed that far from being perfect, Earth could have been much better in ways that should not have been difficult to create for an omnipotent god capable of creating a universe [71]. They also showed that these planets could be relatively common in this universe.

The simple facts and inevitability of evolution are perfectly adequate for explaining why life appears to be more or less perfectly adapted to the present conditions on Earth, without the need to invoke magic. Living things being very well suited to the conditions in which they evolved is not at all the same as those conditions being perfect for living things to live in. Evolution can mould a species to the environment but it can do nothing at all about the planet itself.

Evolution cannot influence gravity or the magnetic field of the planet, or what meteorites might smash into it. Nor can it affect the process of plate tectonics or the rate at which the sun is burning up its fuel, nor what will become of the planet when the sun blows up.

In fact, evolution of life can only influence the evolution of environments to a limited extent and none of it can be planned and made the purpose of evolution. So Earth has many environments such as arid deserts, arctic icecaps, permafrost, snow-covered mountain ranges and deep ocean trenches where only highly specialised organisms can exist and where life, for the most part, is a rarity and where life for most of Earth's species would be impossible.

But, a planet perfect for life would be one where life was abundant everywhere, where the planet had a longer life in which evolved life could flourish and mature. The environment would be such that basic maintenance activities like respiration, body temperature stability, etc. took minimal effort, and the planet would be protected from meteorites, comets and dangerous solar radiation.

The Unintelligent Designer

René Heller and colleagues analysed a range of factors which could have produced a better planet for life. For example, Earth is not in the centre of the habitable zone around the sun as creationists like to pretend. In fact it is close to the inner edge of this zone. As the sun gets hotter, in the next 1-2 billion years, Earth will become a hot, uninhabitable rock like Venus. Life has been on Earth now for about 3.5 billion years so we are already well beyond halfway to heat-death extinction even if Earth manages to survive a runaway greenhouse effect.

Had Earth been a little larger and orbiting an orange star it could all have been so much better. The star would be around for about 7-10 billion years instead of the 5-6 billion we get. The higher gravity would have meant shallower seas and low, rounded hills on chains of islands in warm seas - exactly the conditions which have the highest biodiversity on Earth.

A slightly higher gravity would have made the atmosphere a little denser, so it would have been easier to take in oxygen and respiratory systems could have been smaller and taken less effort. Flight would have been easier for larger creatures. It would also have increased plate tectonic activity without dramatic and destructive earthquakes, tsunamis, volcanoes and upheavals of mountain-building that Earth has experienced. This, in turn would have increased the core temperature and the magnetic field, so helping to deflect dangerous solar radiation.

We already know that the orange star, Alpha Centauri B, our Sun's nearest neighbour, hosts a rocky planet which is already 6 billion years old. This planet is probably too close to its sun to host life but we are now discovering that exoplanets (i.e. planets outside our solar system) are common. There were clearly very many far better planets creationists' supposed creator could have put us on.

But then, the existence of these other suns with their planetary system was unknown to the authors of the holy books.

So where then can we see the evidence of this intelligent design if even the planet it put it all on wasn't very well designed? How was a poorly designed planet intended to make it easier for humans to worship gods and obey all their rules?

But it gets far worse.

Not only is this demonstrably **not** a perfect planet but much of it seems to be the product of an evil designer, if we accept for the sake of argument that there is design. A great deal of the complexity we see in all living organisms has come about because the environment tends to be hostile, so organisms constantly need to adapt or die. A great deal of the complexity is necessary because other life–forms are part of the hostile environment, living on, in or off other species as parasites or predators.

For almost all higher life, much of which is sentient and will feel pain, life, far from being some romantic idyll, is nasty, brutish and short. For most members of higher species, there is no such thing as a peaceful death. Death is almost invariably painful; either caused by being killed and eaten, by disease and ill–health or by starvation due to age and infirmity, if a predator doesn't find them first.

Higher animals die in pain. There can be little doubt about that. Put simply, pain tells us something is wrong. Pain draws our attention to injury or disease. Pain says do something or don't do something; guard me, rest me or don't use me. Don't walk on that broken ankle because it needs to be rested. Don't carry on with that chest pain but slow down and take a rest. Don't bite on that tooth or raise that broken arm. Close your eyes and sleep when that headache becomes unbearable. Put your hand over your ear when cold wind makes it ache and change your shoes when that blister bursts...

Pain even initiates reflexes which happen before our brains have noticed. These spinal reflexes have evolved to protect various parts of our bodies and pain is the signal to act automatically without the normal luxury of thinking about it first.

Pain has evolved as a signal. It is unpleasant because that tells us to try to stop it by resting or guarding the hurting part of our body. Pain is unpleasant because we have evolved to perceive it as unpleasant. Being unpleasant means we do something about it to reduce the unpleasant sensation.

Consider a patient dying in extreme pain of cancer, or an abscess, or a disabling injury in the absence of any pain relief? What possible purpose could that serve the individual? Consider a gazelle dying of the shock of having its intestines pulled out and its liver eaten by lions whilst still alive, or the zebra having a leg torn off by a crocodile as it is slowly drowned. How does pain serve these individual?

Nature is unemotional and entirely lacking in compassion. Nature doesn't care about the suffering of a prey species as it is eaten and yet we can be quite sure that every sentient creature, and probably many others, feels pain. Nature has no concern at all for the discomfort or distress of an animal suffering from infection or dying of disease or simply starving to death of old age.

The fate of almost every living multi-cellular thing is to die of disease, or by being eaten, or of starvation due to injury or old age. There are very many ways to die and none of them are pleasant. Millions of feeling animals die every day in great pain. A system which has evolved to keep you alive is useless when you are dying, and yet it is still demanding you do something even when there is nothing you can do.

So why should we have evolved something we don't like and why would it be at its most insistent when at its most useless? What intelligent designer would design such a thing?

Because evolution isn't driven by what we like or dislike; evolution is driven by whatever ensures we have more descendants than we would otherwise have. Evolution is determined by what is in the interests of our genes because it is our genes which either survive in the next generation, or don't. And there is no benefit to our genes in evolving a mechanism to turn pain off when it is no longer any use.

So, evolution has provided us with something we don't like, and this is perfectly understandable in terms of mindless, unemotional, uncaring, genetic evolution.

What is not understandable is how this could have been designed by an intelligent, loving, caring and compassionate god. If pain has been designed by a god then that god must be a stupid, cruel, sadistic and hateful god.

If the ultimate purpose of some reputed intelligent designer was to design human life so that it would worship and obey it and the rest of 'creation' to serve mankind, and if that designer had been a maximally good god then it could not possibly have designed the world we live in.

There is no evidence of intelligence in design; all there is, is evidence of unintelligent, undirected, unemotional, purposeless design. Life is unnecessarily complex, prolifically wasteful and has no ultimate purpose. Life is not intelligently designed; it is moulded by a process that gives a superficial resemblance of design to those who remain stoically ignorant of the details.

Intelligent Design is a hoax perpetrated against a scientifically illiterate and credulous people for political objectives.

The Unintelligent Designer

Bibliography

1. **Matsumura, Molleen and Mead, Louise.** Ten Major Court Cases about Evolution and Creationism. *National Center for Science Education.* [Online] https://ncse.com/library-resource/ten-major-court-cases-evolution-creationism.

2. **Hunter, Mat.** What is Design and Why It Matters. *CiC UK To The World.* [Online] 2014. http://www.thecreativeindustries.co.uk/uk-creative-overview/news-and-views/view-what-is-design-and-why-it-matters.

3. **Spacey, John.** Design: Complexity vs Simplicity. *Simplicable.* [Online] 2 June 2016. https://simplicable.com/new/complexity-vs-simplicity.

4. **Rubicondior, Rosa.** Intelligent Design - What A Lot Of Balls! *Rosa Rubicondior.* [Online] 8 February 2014. [Cited: 12 July 2018.] http://rosarubicondior.blogspot.com/2014/02/intelligent-design-what-lot-of-balls.html.

5. —. The Teleological Fallacy or Paley's Broken Watch. *Rosa Rubicondior.* [Online] 27 January 2012. [Cited: 18 July 2018.] http://rosarubicondior.blogspot.com/2012/01/teleological-fallacy-or-paleys-broken.html.

6. —. *The Light of Reason: Atheism, Science and Evolution.* 1. Oxford : CreateSpace, 2015. pp. 135-138. Vol. 2. ISBN-13: 978-1517105181.

7. *The Natural History of Ascorbic Acid in the Evolution of the Mammals and Primates and Its Significance for Present Day Man.* **Stone, Irwin.** 1 & 2, 1972, Orthomolecular Psychiatry, Vol. 1 , pp. 82-89. Posted online 2013.

8. *Evolution and the Need for Ascorbic Acid.* **Pauling, Linus.** 4, 15 December 1970, Proceedings of the National Acaemy of Science of the United States of America, Vol. 67, pp. 1643–1648.

9. **Rational Wiki.** Laryngeal nerve. *Rational Wiki.* [Online] 8 June 2018. [Cited: 19 July 2018.] https://rationalwiki.org/wiki/Laryngeal_nerve.

10. **Berkeley University.** The arms race. *Understanding Evolution.* [Online] [Cited: 12 July 2018.] https://evolution.berkeley.edu/evolibrary/article/armsrace_01.

11. *The butterfly plant arms-race escalated by gene and genome duplications.* **Edger, Patrick P., et al., et al.** s.l. : Proceedings of the National Academy of Sciences (PNAS), 22 June 2015.

12. **Meissen, Roger.** Scientists discover how caterpillars created condiments. *Decoding Science.* [Online] Bond Life Sciences Center at the University of Missouri, 22 June 2015. [Cited: 12 July 2018.] https://decodingscience.missouri.edu/2015/06/22/scientists-uncover-how-caterpillars-created-condiments/.

13. *The king cobra genome reveals dynamic gene evolution and adaptation in the snake venom system.* **Vonk, Freek J., et al., et al.** 110, 17 December 2013, Proceedings of the National Academy of Sciences (PNAS), Vol. 51, pp. 20651-20656. ISSN: 1091-6490.

14. **Holmes, Bob.** Extreme evolution: How snakes became the über-eater. *New Scientist.* [Online] 7 June 2014. [Cited: 12 July 2018.] https://www.newscientist.com/article/mg22229720-700-extreme-evolution-how-snakes-became-the-uber-eater/.

15. **Rubicondior, Rosa.** Snake Bite Shock for Creationism. *Rosa Rubicondior.* [Online] 2014 June 2014. [Cited: 12 July 2018.] http://rosarubicondior.blogspot.com/2014/06/snake-bite-shock-for-creationism.html.

16. *Catch a tiger snake by its tail: Differential toxicity, co-factor dependence and antivenom efficacy in a procoagulant clade of Australian venomous snakes.* **Lister, Callum, et al., et al.** s.l. : Elsevir, 27 November 2017, Comparative Biochemistry and Physiology Part C: Toxicology & Pharmacology, Vol. 202, pp. 39-54.

17. *Ungulate saliva inhibits a grass–endophyte mutualism.* **Tanentzap, Andrew J., Vicari, Mark and Bazely, Dawn R.** s.l. : Royal Society Publishing, 23 July 2014, Biology Letters.

18. *Plants Can Benefit from Herbivory: Stimulatory Effects of Sheep Saliva on Growth of Leymus chinensis.* **Liu, Jushan , et al., et al.** 1, s.l. : PLOS Biology, 3 January 2012, PLoS ONE, Vol. 7. e29259.

19. **Jacobs, David.** Explainer: the evolutionary arms race between bats and moths. *The Conversation UK.* [Online] 26 June 2015. [Cited: 13 July 2018.] http://theconversation.com/explainer-the-evolutionary-arms-race-between-bats-and-moths-43675.

20. *The evolutionary origins of beneficial alleles during the repeated adaptation of garter snakes to deadly prey.* **Feldman, Chris R., et al., et al.** 106, 11 August 2009, Proceedings of the National Academy of Sciences (PNAS), Vol. 32, pp. 13415-13420.

21. *Genetic architecture of a feeding adaptation: garter snake (Thamnophis) resistance to tetrodotoxin bearing prey.* **Feldman, Chris R., et al., et al.** 277, 3 June 2010, Proceedings of the Royal Society B, Vol. 2010, pp. 3317-3325.

22. **Creation Minestries International.** Why doesn't Sir David Attenborough give credit to God? *Creation.com.* [Online] [Cited: 14 July 2018.] https://creation.com/why-doesnt-sir-david-attenborough-give-credit-to-god.

23. *Tales from the crypt: a parasitoid manipulates the behaviour of its parasite host.* **Weinersmith, Kelly L. , et al., et al.** 1847, s.l. : The

Royal Society, 25 January 2017, Proceedings of the Royal Society B, Vol. 284.

24. *Who is the puppet master? Replication of a parasitic wasp-associated virus correlates with host behaviour manipulation.* **Dheilly, Nolwenn M. , et al., et al.** 1803, s.l. : The Royal Society, 11 February 2015, Proceedings of the Royal Society B, Vol. 282. ISSN: 1471-2954.

25. *Studies on the biology of Dicrocoelium dendriticum (Rudolphi, 1819) Looss, 1899 (Trematoda: Dicrocoeliidae), including its relation to the intermediate host, Cionella lubrica (Müller). III. Observations on the slimeballs of Dicrocoelium dendriticum.* **Krull, W. H. and Mapes, C. R.** 2, April 1952, Cornell Veterinarian, Vol. 42, pp. 253-76. PMID: 14926337.

26. **Petruzzello, Melissa.** Pilobolus. *Encyclopaedia Britannica.* [Online] Encyclopaedia Britannica, 30 July 2013. [Cited: 15 July 2018.] https://www.britannica.com/science/Pilobolus-fungus-genus.

27. **World Health Organization.** Schistosomiasis. *World Health Organization Fact Sheet.* [Online] 20 February 2018. [Cited: 14 July 2018.] http://www.who.int/news-room/fact-sheets/detail/schistosomiasis.

28. *Onchocerciasis: the Role of Wolbachia Bacterial Endosymbionts in Parasite Biology, Disease Pathogenesis, and Treatment.* **Tamarozzi, Francesca, et al., et al.** 3, Washington DC : American Society for Microbiology, 1 Juky 2011, Clinical Microbiology Reviews, Vol. 24, pp. 459–468.

29. **World Health Organization.** Life-cycle of Onchocerca volvulus. *African Programme for Onchocerciasis Control (APOC).* [Online] World Health Organization , 2018. [Cited: 16 July 2018.] http://www.who.int/apoc/onchocerciasis/lifecycle/en/.

30. **Rubicondior, Rosa.** Creationists, Flying In The Face Of Reason. *Rosa Rubicondior.* [Online] 9 February 2013. [Cited: 16 July 2018.] http://rosarubicondior.blogspot.com/2013/02/creationists-flying-in-face-of-reason.html.

31. **Steverding, Dietmar.** The history of African trypanosomiasis. *Parasites & Vectors.* [Online] BioMed Central (BMC), 12 February 2008. [Cited: 16 July 2018.] https://parasitesandvectors.biomedcentral.com/articles/10.1186/1756-3305-1-3.

32. **Behe, Michael J.** *Darwin's Black Box : The Biochemical Challenge to Evolution.* s.l. : Free Press, 1996. ISBN-13: 978-0684827544.

33. **Justia US Law.** *Kitzmiller v. Dover Area School Dist., 400 F. Supp. 2d 707 (M.D. Pa. 2005).* 04cv2688., s.l. : Middle District of Pennsylvania, 20 December 2005.

34. *TSETSE GENETICS: Contributions to Biology, Systematics, and Control of Tsetse Flies.* **Gooding, R.H. and Krafsur, E.S.** 1, 2005, Annual Review of Entomology, Vol. 50, pp. 101-123 .

35. **Behe, Michael J.** *The Edge of Evolution : The Search for the Limits of Darwinism.* s.l. : Free Press, 2007. ISBN-10: 0743296206.

36. **Miller, Kenneth R.** Falling over the edge. *Nature.* 28 June 2007, 447, pp. 1055–1056.

37. **World Health Organization.** Malaria. *World Health organization.* [Online] 2018. [Cited: 16 July 2018.] http://www.who.int/malaria/en/.

38. *Trichomoniasis.* **Schwebke, Jane R. and Burgess, Donald.** 4, s.l. : American Society for Microbiology, 1 October 2004, Clinical Microbiology Reviews, Vol. 17, pp. 794-803.

39. *An Expanded Inventory of Conserved Meiotic Genes Provides Evidence for Sex in Trichomonas vaginalis.* **Malik, Shehre-Banoo, et al., et al.** 8, 6 August 2008, PLoS ONE, Vol. 3. e2879.

40. **Medical XPress.** Scientists crack the genome of the parasite causing trichomoniasis. *Medical XPress.* [Online] 12 January 2007. [Cited: 17 July 2018.] https://medicalxpress.com/news/2007-01-scientists-genome-parasite-trichomoniasis.html.

41. **Harvard Medical School.** When parasites catch viruses. *PhyOrg.* [Online] 7 November 2012. [Cited: 17 July 2018.] https://phys.org/news/2012-11-parasites-viruses.html.

42. **Rosenau, Joshua.** The Wedge Document. *NCSE.com.* [Online] National Centre for Science Education, 14 October 2008. [Cited: 19 July 2018.] https://ncse.com/creationism/general/wedge-document.

43. *Ubiquity and Diversity of Human-Associated Demodex Mites.* **Thoemmes, Megan S., et al., et al.** 8, 27 August 2014, PLoS ONE, Vol. 9. e106265.

44. **Rubicondior, Rosa.** Seeing Eye to Eye With a Butterfly. *Rosa Rubicondior.* [Online] 23 April 2012. [Cited: 20 July 2018.] http://rosarubicondior.blogspot.com/2012/04/seeing-eye-to-eye-with-butterfly.html.

45. *Contributions to an insect fauna of the Amazon Valley. Lepidoptera: Heliconidae.* **Bates, W. H.** London : s.n., 1862, Transactions of the Linnean Society of London., Vol. 23, pp. 495-566.

46. **Rubicondior, Rosa.** *What Makes You So Special? : From the Big Bang to You.* Oxford : CreateSpace Independent Publishing Platform, 2017. pp. 64-65. ISBN-13: 978-1546788294.

47. **Lane, Nick and Le Page, Michael.** How life evolved: 10 steps to the first cells. *New Scientist.* 14 October 2009.

48. **Dawkins, Richard.** *Unweaving the Rainbow.* London : Penguin Books, 1998. pp. 12-13. ISBN-10: 0140264086.

49. *On the origin of mitosing cells.* **Sagan, Lynn.** 3, s.l. : Elsevier, March 1967, Journal of Theoretical Biology, Vol. 14, pp. 225-274.

50. **Lane, Nick.** *Life Ascending: The Ten Great Inventions of Evolution.* Paperback. s.l. : Profile Books, 2010. ISBN-10: 1861978189.

51. *Catalytic by-product formation and ligand binding by ribulose bisphosphate carboxylases from different phylogenies.* **Pearce, F. Grant.** 3, London : Portland Press, 1 November 2006, Biochemical Journal, Vol. 399, pp. 525-534.

52. *Human endogenous retroviruses: transposable elements with potential?* **Nelson, P. N., et al., et al.** 1, 31 August 2004, Clinical & Experimental Immunology, Vol. 138, pp. 1-9.

53. **Answers in Genesis.** DNA Prevents Reproducing Between Two Species. *Answers in Genesis.* [Online] AiG, 31 October 2009. [Cited: 25 July 2018.] https://answersingenesis.org/genetics/junk-dna/dna-prevents-reproduction/.

54. *Species-Specific Heterochromatin Prevents Mitotic Chromosome Segregation to Cause Hybrid Lethality in Drosophila.* **Ferree, Patrick M. and Barbash, Daniel A. .** 7, 27 October 2009, PLoS Biology, Vol. 10. e1000234.

55. *Is junk DNA bunk? A critique of ENCODE.* **Doolittle, W. Ford.** 14, 2 April 2013, Proceedings of the National Academy of Sciences (PNAS), Vol. 110, pp. 5294-5300.

56. *An integrated encyclopedia of DNA elements in the human genome.* **Consortium, The ENCODE.** 6 September 2012, Nature, Vol. 489, pp. 57–74.

57. **Luskin, Casey.** Junk No More: ENCODE Project Nature Paper Finds "Biochemical Functions for 80% of the Genome". *Evolution News.* [Online] Discovery Institute, 5 September 2012. [Cited: 26 July 2018.] https://evolutionnews.org/2012/09/junk_no_more_en_1/.

58. **Tomkins, Jeffrey P.** Junk DNA Myth Continues Its Demise. *Institute for Creation Research.* [Online] 31 October 2012. [Cited: 26 July 2018.] http://www.icr.org/article/junk-dna-myth-continues-its-demise/.

59. **Le Page, Michael.** At least 75 per cent of our DNA really is useless junk after all. *New Scientist.* 17 July 2017.

60. *Architecture and evolution of a minute plant genome.* **Ibarra-Laclette, Enrique, et al., et al.** 6 June 2013, Nature, Vol. 498, pp. 94–98.

61. *The Norway spruce genome sequence and conifer genome evolution.* **Nystedt, Björn, et al., et al.** 22 May 2013, Nature, Vol. 497, pp. pages 579–584.

62. **Rubicondior, Rosa.** Christmas Tree Tease for Creationists. *Rosa Rubicondior.* [Online] 15 July 2014. [Cited: 22 July 2018.] http://rosarubicondior.blogspot.com/2014/07/christmas-tree-tease-for-creationists.html.

63. **Carey, Nessa.** *The Epigenetics Revolution: How Modern Biology is Rewriting Our Understanding of Genetics, Disease and Inheritance.* s.l. : Icon Books, 2012. ISBN-10: 1848313470.

64. *Persistent epigenetic differences associated with prenatal exposure to famine in humans.* **Heijmans, Bastiaan T., et al., et al.** 44, 4 November 2008, Proceedings of the National Academy of Sciences (PNAS), Vol. 105, pp. 17046-17049.

65. **Hall, Brian K.** Developmental Mechanisms Underlying the Formation of Atavism. *Biological Reviews.* February 1984, Vol. 59, 1, pp. 89-122.

66. *Developmental Basis of Phallus Reduction during Bird Evolution.* **Herrera, Ana M., et al., et al.** 12, s.l. : Elsevier, 6 June 2013, Current Biology, Vol. 23, pp. 1065-1074.

67. *A molecular mechanism for the origin of a key evolutionary innovation, the bird beak and palate, revealed by an integrative approach to major transitions in vertebrate history.* **Bhullar, B. S., et al., et al.** 7, s.l. : John Wiley & Sons, Inc., 12 May 2015, Evolution, Vol. 69, pp. 1665-1677.

68. *Progressive Loss of Function in a Limb Enhancer during Snake Evolution.* **Kvon, Evgeny Z., et al., et al.** 3, s.l. : Elsevier, 20 October 2016, Cell, Vol. 167, pp. 633-642.

69. *Loss and Re-emergence of Legs in Snakes by Modular Evolution of Sonic hedgehog and HOXD Enhancers.* **Leal, Francisca and Cohn, Martin J.** 21, s.l. : Elsevier, 20 October 2016, Current Biology, Vol. 26, pp. 2966 - 2973.

70. *Human tails and pseudotails.* **Dao, Anh H. and Netsky, Martin G.** 5, s.l. : Elsevier, 1 May 1984, Human Pathology, Vol. 15, pp. 449-453.

71. *Superhabitable Worlds.* **Heller, René and Armstrong, John.** 1, s.l. : Mary Ann Liebert, Inc, 16 January 2014, Astrobiology, Vol. 14, pp. 50-66.

The Unintelligent Designer

Index

Index

Index

Books by Rosa Rubicondior

The Light of Reason Series:

The Light of Reason: And Other Atheist Writings.
Irreverent essays, thought-provoking articles and humorous items on atheism, religion, science, evolution, creationism and related issues.

(Hardcover|) ISBN-13: 979-8512173916	£13.50 (US $18.50)
(Paperback) ISBN-10: 1516906888, ISBN-13: 978-1516906888	£9.95 (US $14.95)
(Kindle) ASIN: B014N0IPVI	£3.95 (US $5.99)

The Light of Reason: Volume II – Atheism, Science and Evolution.
Thought-provoking essays on the conflict between fundamentalist religion and science, and exposing the anti-science, extremist political agenda of the modern creationist industry.

(Hardcover) ISBN-13: 979-8512191040	£13.50 (US $18.50)
(Paperback) ISBN-10: 1517105188, ISBN-13: 978-1517105181	£9.95 (US $14.95)
(Kindle) ASIN: B014N0IR16	£3.99 (US $5.99)

The Light of Reason: Volume III – Apologetics, Fallacies, and Other Frauds.
Thought-provoking essays and articles on religion and atheism, dealing with religious apologetics, fallacies, miracles and other frauds

(Hardcover) SBN-13: 979-8512526002	£12.00 (US $17.00)
(Paperback) ISBN-10: 151710761X, ISBN-13: 978-1517107611	£6.95 (US $9.95)
(Kindle) ASIN: B014N0IRE8	£2.99 (US $3.99)

The Light of Reason: Volume IV - The Silly Bible.
Exposing the absurdities, contradictions and historical inaccuracies in the Bible and advancing the case for atheism and against religion. This volume, the fourth in the Light of Reason series, deals with contradictions and absurdities in the Bible.

(Hardcover) ISBN-13: 979-8512539392	£13.50 (US $18.50)
(Paperback) ISBN-10: 1517108209, ISBN-13: 978-1517108205	£8.95 (US $13.95)
(Kindle) ASIN: B014N0IR8E	£3.99 (US $4.99)

The Light of Reason: And Other Atheist Writing. (all 4 volumes in one e-book)
Based on the Rosa Rubicondior science and Atheism blog, this is a collection of Atheist and science articles, some short, others lengthier, exploring the interface between religion and science and which have been published over some four years.

(Kindle only) ASIN: B013DYOK32	£6.34 (US $9.95)

147

The Unintelligent Designer

Other books on science, Atheism and theology

An Unprejudiced Mind: Atheism, Science & Reason.

Essays on science and theology from a scientific atheist perspective, exploring particularly evolution versus creationism.

(Hardcover) ISBN-13: 979-8512554685	£13.10 (US $18.50)
(Paperback) ISBN-10: 1522925805, ISBN-13: 978-1522925804	£9.95 (US $14.95)
(Kindle) ASIN: B019UGXPM4	£3.99 (US $5.95)

Ten Reasons To Lose Faith: And Why You Are Better Off Without It.

Why faith is not only a fallacy and useless as a route to the truth but is actually harmful to society and to the individual. It systematically dismantles the standard religious apologetics and shows them to be bogus and deliberately constructed to mislead.

(Hardcover) ISBN-13: 979-8509108433	£16.00 (US $22.00)
(Paperback). ISBN-13:978-1530431953, ISBN–10: 1530431956	£10.75 (US $14.75)
(Kindle) ASIN: B01DGVO3JS	£6.95 (US $8.95)

What Makes You So Special? : From the Big Bang to You.

How did you come to be here, now? This book takes you from the Big Bang to the evolution of modern humans and the history of human cultures

(Hardcover) ISBN-13: 979-8509108433	£13.50 (US $18.00)
(Paperback) ISBN-13: 978-1546788294, ISBN-10: 1546788298	£8.95 (US $11.50)
(Kindle).ASIN: B071FTKXLZ	$6.20 (US $8.25)

The Internet Handbooks series

The Internet Creationists' Handbook: A Joke for the Rest of Us.

A humorous look at creationist apologetics on the Internet, showing the fallacies and dishonest tactics creationists are using to try to recruit scientifically illiterate people into their political cult.

(Paperback),ISBN-13: 978-1721605149, ISBN-10: 1721605149£5.25 (US $7.50)	
(Kindle) ASIN: B07DZF75KD	£3.75 (US $5.00)

The Christian Apologists' Handbook: A Joke for the Rest of Us.

A humorous look at Christian apologetics on the Internet, showing the fallacies and dishonest tactics Christian fundamentalists are using to try to recruit scientifically and theologically illiterate people to their cults, often with political motives.

(Paperback) ISBN-13: 978-1721724727, ISBN–10: 1721724729	£5.25 (US $7.50)
(Kindle) ASIN: B07DYDVMW4	£3.75 (US $5.00)

Books by Rosa Rubicondior

The Muslim Apologists' Handbook: A Joke for the Rest of Us.
A humorous look at Muslim apologetics on the Internet, showing the fallacies and dishonest tactics Muslim fundamentalists are using to try to recruit scientifically and theologically illiterate people to their cuts, often with political motives.

(Paperback) ISBN-13: 978-1721756896, ISBN-10: 1721756892	£5.25 (US $7.50)
(Kindle) ASIN: B07DZF75KD	$3.75 (US $5.00)

The Unintelligent Design Series

The Unintelligent Designer: Refuting the Intelligent Design Hoax

Showing why the superficial appearance of design in living things cannot be attributed to anything like an intelligent designer, as a counter to the politically-motivated Intelligent Design movement.

(Hardcover) ISBN-13: 979-8513528463	£13.50 (US $18.50)
(Paperback) ISBN-10: 1723144215, ISBN-13: 978-1723144219	£9.00 (US $12.50)
(Kindle) ASIN B07G121BMK	£5.00 (US $7.00)

The Malevolent Designer: Why Nature's God is not Good

Showing why, if we accept for the sake of argument the Creationist insistence on Intelligent Design as the best explanation for biodiversity on Earth, the creator god they purport to worship could not be regarded as anything other than a malevolent evil, assiduously designing suffering into its creation as though it hates it and wants it to suffer in unimaginably horrible ways.

Illustrated by Catherine Hounslow-Webber

(Hardcover) ISBN-13: 979-8511295442	£15.00 (US $18.00)
(Paperback) SBN-13; 979-8670361729	£9.10 (US $12.50)
(Kindle) ASIN: B08L9S8F5F	£5.50 (US $7.60)

Publish under the name Bill Hounslow – Oxfordshire Childhood series.

In The Blink of an Eye: Growing Up in Rural Oxfordshire

A frank recollections of life as feral children in the small North Oxfordshire hamlet of Fawler during the 1950s and 60s, on the brink of major change as we approached the television age and the final stages in the domestication of children was about to begin.

Additional material by Patricia Broome

(Hardcover) ISBN-13: 979-8511967400	£13.00 (US $18.50)
(Paperback) ISBN-10: 1545350787, ISBN-13: 978-1545350782	£6.50 (US $11.49)
(Kindle) ASIN: B06ZY8JZ92	$6.50 (US $8.95)

The Unintelligent Designer

A Goose for Christmas: Stories from an Oxfordshire Childhood

Slightly imaginative stories, based on real events and people, of childhood adventures in the North Oxfordshire hamlet of Fawler in the 1950s during the post-war austerity, before television, when the children had only what they could get from the woods and fields around them.

Illustrated by Catherine Webber-Hounslow

(Hardcover) ISBN-13: 979-8511907482   £12.75 (US $18.00)
(Paperback) ISBN-13: 978-1981708925, ISBN-10: 1981708928   £8.50 (US $11.50)
(Kindle) ASIN: B07GFJ85P8

www.ingramcontent.com/pod-product-compliance
Lightning Source LLC
Chambersburg PA
CBHW071309220526
45468CB00001B/306